流光溢彩 THE COLORS LIGHTING DESIGN
灯光设计

欧朋文化 策划　黄滢 马勇 主编

中国·武汉

目录 CONTENTS

A 室外灯光 OUTDOOR LIGHTING

006	柏林灯光展	Berlin Lighting Exhibition
012	莫斯科"光之环"灯展	Moscow International Festival "Circle of Light"
016	里昂灯光节	Lyon Lighting Festival
022	"神秘博士"别墅	The House of Doctor Who
026	悉尼歌剧院灯光展	The Sydney Opera House Lighting Exhibition
034	中国·哈尔滨国际冰雪节	Harbin International Snow and Ice Festival
042	悉尼海港大桥灯光秀	Sydney Harbor Bridge Lighting Show
044	巴库水晶大厅	Baku Crystal Hall
046	华沙国家体育馆	National Stadium in Warsaw
050	迪拜亚特兰蒂斯酒店开幕灯光秀	The Palm Grand Opening Lights Show, Atlantis
052	筒仓468	Silo 468
056	迪斯尼灯光设计	Disney Lighting

B 室内灯光 INTERIOR LIGHTING

062	食天下酒吧	The World Cuisine, KNOKKE-Zuri
072	鸡尾酒酒吧	The Smokehouse Room—Cocktail Bar
078	罗克西	Roxy
086	洞穴休息酒吧	Katakomb Lounge Bar
092	夜飞	Night Flight Club
100	高低会所	Smoking Club Hi/Lo
106	莆堡酒吧	POPO Club
112	马来西亚SOJU 吉隆坡店	SOJU
116	南京Enzo酒吧	Enzo Bar, Nanjing
122	B-ONE酒吧会所	B-ONE Lounge Club
132	胖夫人迪斯科舞厅	Fat Lady Discotheque
138	重庆丽芙酒吧	Liv Show Bar

146	Nova 酒吧	Nova
152	江西永修东方玛赛音乐会所	Music Club
162	南通皇后	Club Queen, Nantong
168	广州增城迷笛会酒吧	Mint Bar
174	安徽芜湖星光璀璨娱乐城	Sparkle Entertainment City（Wuhu，Anhui）
178	拉科瓦舞场会所	La Cova Dance Theatre
182	以色列啤酒俱乐部	"Forum Club"－Beer Sheva-Israel
190	大同LOVE 100 酒吧	Love 100 Club
196	重庆CLUB ONE酒吧	Club One, Chongqing
202	台北纯K杭州店	Chun K Party（Hangzhou）
208	皇庭KTV	Royal Courtyard KTV
212	维的雀旗舰店	Videotron's Flagship Store
216	革命酒吧	The Revolution Lounge
222	"八角"会所	Club Octagon
228	吉隆坡Rootz会所	Rootz Club
234	缪特会所	Mute Club
240	烟尘会所	The Smokehouse Room－Nightclub
246	北京麦乐迪KTV月坛店	Melody KTV（Yuetan, Beijing）
254	麦乐迪南京新街口店	Melody KTV（Xinjiekou, Nanjing）
260	广东中山魅力皇爵KTV会所	Charm Baron KTV Club（Zhongshan, Guangdong）
266	咏歌汇2期	Phase 2 of Yongge Hui
272	康业国际会所	Kangye International Club
278	天命夜总会	Kismet Nightclub
282	胜悦国际头皮养护馆台北市南京路店	Shengyue Hair Care（Nanjing East Road, Taibei）
288	Kippo美发沙龙	Kippo Hair and Color Bar
290	意念空间	The Room
292	水族馆酒吧	Aquarium Bar
294	咖啡酒吧	Cafe Bar
296	搜狐玻璃办公室	Glass Office
302	梨泰院华丽丽会所	Glam Lounge
306	黑盒子	Black Box－Dowling Billiards Darts Club
312	克罗维克电影院	Kronverk Cinema

A 室外灯光
OUTDOOR LIGHTING

用灯光为建筑穿上华彩外衣，配合不同风格，变幻万千姿彩。灯光设计师应合理组织夜景照明构架，创造灯光景观环境，塑造夜间视觉焦点，兼顾实用美观与节能环保。

室外灯光设计的八大原则如下：

一、整体性原则

强调现代建筑的整体性，合理规划现代建筑灯光景观的各构成要素；强调群体灯光环境的整体性，处理好单体与群体的关系；重视单体灯光照明与其所处的灯光环境的整体性，协调背景与主体的关系，在强调单个灯具的个性与特色的同时又使之融入区域环境的景观之中。

二、艺术性原则

照明方式、照度、光色等灯光环境要素及其组合的艺术性是灯光景观设计的一个重要目的。照明灯具及其安装的艺术性是灯光景观设计不可忽略的因素。灯具设计的艺术性可以为整个灯光景观设计增添意趣，使美学观赏价值与实用价值紧密结合。

三、开放与创新原则

灯光景观的开放原则是构成现代建筑夜景丰富多彩的基本途径之一。开放原则主要体现在表现手法的开放上，不应简单排斥或一味追求某种灯光表现手法，而应依据具体环境表现主题以及根据实际经济能力来确定表现的方法和采用的设备。创新是灯光景观艺术的灵魂，它既包括创作观念的创新，也包括技术手段的创新，积极运用新工艺、新光源、新材料，特别是应用高新技术，提高景观灯光设施的科技性。

四、时代原则

在设计中，将现代建筑的灯光景观环境与城市发展方向紧密结合起来，现代建筑的灯光总体设计应纳入现代建筑的总体规划成果中去，能随现代建筑总体规划的调整而变化。现代建筑的灯光景观表现形式应符合当代艺术创作的总趋势，在规划中应注意体现灯光景观艺术鲜明的时代特征，体现现代艺术的感染力。在设计中注入活力，追求时代感，追求一种交响乐般的恢弘大气。

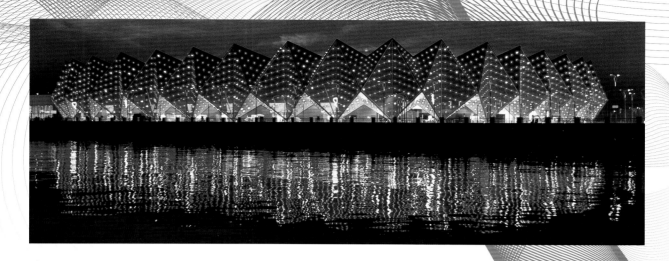

五、个性化与特征原则

通过总体创意、设计方法、设计手段与设计中采用的灯光装置与设备来体现这一原则。使灯光景观设计做到与时代和社会的发展同步。对城市居民行为特征与行为模式的分析是灯光景观设计的重要基础。

六、经济实用原则

统筹处理使用寿命、维修管理费用和一次性投资的关系，使灯光景观规划做到经济合理；保证各区域功能和活动所需的照度水平，满足视觉要求；考虑照明灯具的使用和维护方便。

七、安全与环保原则

现代建筑灯光设计的安全原则包括灯光系统自身的安全性与灯光环境对社会治安的促进作用；充分利用高科技，开发新的高效、节能的照明手段；综合考虑各种环境因素，选择经济、实用、节能的电光源；合理选择灯具安装位置，使照明灯具发挥最大的照明效率；控制眩光，防止光污染。

八、建筑与周边环境融合的原则

建筑并不是一个孤立的个体，而是与周边的总体环境与建筑相融合。灯光景观的设计也应当考虑与周边环境灯光景观的协调与呼应。

资料参考：百度文库

流光溢彩 灯光设计
THE COLORS LIGHTING DESIGN

柏林灯光展
BERLIN LIGHTING EXHIBITION

"柏林灯展"源于2005年，由在柏林出生的比吉特·赞德发起。自第一届起，该灯展发展兴盛，成了远近闻名的灯展之一。2012年柏林灯光展位于布兰登堡门的部分，是该届灯展的主体部分。3D的宏伟设施，4个高清晰照明沿布兰登堡门的前面洒下满天的光辉，直至巴黎广场。布兰登堡门的平面，复杂的外观，众多的台阶、角落要想在夜间得到充分的展示，必须有强有力的明亮灯光。所使用的松下PT-DZ21K投影仪，拥有2万流明的亮度，并且可以在车上自由调节。而且该投影仪体格小，体重轻，16A的标准功率便于安装、携带。

即使停电，有了富有创意的4头灯组，也不用担心投影仪的运作。来自柏林的TST GmbH租赁公司，负责设备的提供及"布兰登堡门"前的PT-DZ21K投影仪的安装。同时，该公司与AV Stumpfl Wings Platinum携手积极联系媒体，并于整个灯展期间对安装工作全程管理。2012年的柏林灯展，PT-DZ21K为世人留下了全新视觉体验。

到了2013年，柏林灯展则以"重新发现现代艺术"为主题。很多大牌艺术家齐聚本届盛会，为其成功集思广益。现代的摩天大楼，甚至其周围地面都变得流光溢彩，精彩纷呈。

柏林广播塔穿上了粉色的外衣，是对10月11日"联合国国际女孩日"的礼赞。波茨坦广场的玻璃装饰板，高达8米。夜晚时分，闪着红光。夏洛腾堡宫汇集了18世纪法国的名画。而柏林美国大使馆的灯展以"我是柏林人"、"我有一个梦想"为主题。

另外，议会大厦、柏林大教堂等知名地标性建筑都参与了本次柏林灯展。

室外灯光
OUTDOOR LIGHTING 007

008 流光溢彩 灯光设计
THE COLORS LIGHTING DESIGN

The Berlin Festival of Lights was founded in 2005 by the Berlin-born Birgit Zande. From there it developed into one of the biggest and best known public illumination festivals. For this year the installation at the Brandenburg Gate was developed into the flagship attraction of the festival. Zander & Partner with its technical director of the festival, Gunter Birnbaum, developed and produced a stunning 3D video mapping application where four high brightness projectors are shooting their light to the front side of the Brandenburg Gate from the Pariser Platz. To illuminate a surface of the size of the Brandenburg Gate, even at night, requires a strong and bright projection technology. At the same time the complex shape of the building with its multiple steps and corners does not provide an even projection surface, so that the projection content requires geometric adjustment. The Panasonic PT-DZ21K projector looked the ideal projector for this, a projector with 20,000 Lumen brightness and onboard geometric adjustment. Additionally with its compact size, low weight and 16A standard power supply it that eased the setup and handling tremendously.

With the innovative four-lamp system the projector provided a high failure safety. This was seen as essential for an installation with such a strong exposure. The Berlin based Rental Company TST GmbH served as the equipment supplier to the festival and as that installed four PT-DZ21K in front of the Brandenburg Gate, connected to a media server with AV Stumpfl Wings Platinum and managed this installation throughout the festival period. Resulting in one of the most stunning and eye-catching

installations at 2012 Berlin Festival of Lights, Panasonic PT-DZ21K proved once again it's superior performance in the events sector, creating a new visual experience at the Brandenburg Gate. For twelve days this October, Berlin will be brightly illuminated with splashes of light that transform the city into a colorful work of art. Berlin's ninth annual Festival of Lights is one of the largest light shows in the world: more than 100 buildings, streets and landmarks will be brightly decorated with eye-catching patterns and images.

This year's festival is titled "Rediscovering Modern Art"— a theme that will be fulfilled with light projections of famous paintings and ideas by artists such as Roy Lichtenstein, Andy Warhol, Piet Mondrian and Keith Haring. Quotes by modern artists will creep across buildings, bright lights will make skyscrapers twice as magnificent and even the ground will be illuminated with brilliant splashes of color.

010 流光溢彩 灯光设计
THE COLORS LIGHTING DESIGN

The Berlin Funkturm will be illuminated in pink, thereby honoring the United Nations' International Day of the Girl, which takes place on Oct. 11. Potsdamer Platz will be decorated with 8-meter- high (26 feet) blades of grass with tips that glow red at night. A collection of 18th century French paintings will illuminate the Charlottenburg Palace, while the US Embassy in Berlin will feature light shows based on two motifs: "Ich bin ein Berliner" and "I have a dream".

Other illuminated landmarks include the Reichstag, the Victory Column, the Berlin Cathedral, the Fernsehturm (TV Tower) and the Brandenburg Gate.

MOSCOW INTERNATIONAL FESTIVAL "CIRCLE OF LIGHT"

流光溢彩 灯光设计

莫斯科每年一度的灯展"光环"可谓是年之盛典。届时,设计师、专家云集。借助于2D、3D的形式,为莫斯科地标建筑披上光华的外衣,为世人呈上光的盛宴。莫斯科也如同一块大画布,供各路媒体和各种光之造型一展身手。

2014年灯展的主题是"环绕光的路途"。灯展期间,还有以"视觉艺术"为主题的视频比赛,来自于26个国家的专业人士齐聚一堂,交流借鉴。

闭幕式期间,令人激动的3D展现带领世人穿越,回到遥远的埃及、希腊。文艺复兴时期和中世纪的风情也得到了展示。诸如"自然"、"芭蕾"等为莫斯科人喜闻乐见的主题在本届灯光展上也得到了尽情展现。

今年是莫斯科灯展的第四届,规模盛况超出了以往。如莫斯科大剧院、察里津博物馆、驯马场广场等地点都参与了本届灯展的举办。四天灯展期间,有超过600万民众参观。另外,每天还有50万网民登录网站,收看网上视频。

The Moscow international festival "Circle of Light" is yearly spectacle, during which light designers and specialists in 2D-and 3D-graphics use the architectural space of Moscow as a canvas for multimedia and luminance installations.

The main theme for this years' event was "Journey Round the Light". The festival included a spectacular video mapping contest called "Art Vision", where professional and amateur video mappers from 26 countries took part. During the closing ceremony, an impressive 3D-mapping installation immersed its captive audience into the exciting history theatre during Ancient Egypt and Greece, the Renaissance and the middle Ages. Some of the other themes presented to the lucky Muscovites were nature and the ballet.

This, the fourth festival, became the largest in its history. The locations for the event, Ostankino, the VDNKh, the Bolshoi Theater, the Tsaritsyno museum and Manezhnaya square all hosted massive light displays. During the four days of the festival, more than six million people visited the various locations and more than 500 thousand people visited lightfest.ru website every day.

室外灯光 OUTDOOR LIGHTING | 013

里昂灯光节
LYON LIGHTING FESTIVAL

流光溢彩 灯光设计
COLORS LIGHTING DESIGN

灯光艺术节是法国里昂市的一个宗教节日，每年12月8日开始，一般持续4天。里昂的每户居民都会在自家窗户外摆放点着的蜡烛，以营造满街烛光的效果。同时，城中的宗教场所或一些公共场所也会摆放蜡烛，其中最有名的地点包括富维耶圣母院和沃土广场。

但据史料记载，"里昂灯光节"的由来颇具传奇色彩。1852年，里昂人重建大教堂的钟楼，并重塑圣母马利亚的镀金铜像。原计划在当年9月8日（即圣母诞生日）举行纪念活动，但是索恩河突然涨水，纪念活动不得不推迟到12月8日。孰知12月8日上午，里昂下了一场特大暴雨。正当大主教准备再次宣布推迟纪念活动，里昂人也变得垂头丧气时，大雨在傍晚时分终于停下来，天空完全放晴。兴奋的人们纷纷在窗前点起蜡烛，并走上街头载歌载舞。满城的烛光直到次日天明才渐渐熄灭。自此，12月8日就正式定为"里昂灯光节"的开始日。

灯光节是里昂最大的城市盛会之一，随着时代的发展，这一传统节庆的内容不断丰富，影响力不断扩大，参与范围从里昂扩展到全法国乃至全球，逐渐成为里昂的新名片。值得注意的是，里昂灯光节设计一直采用节能照明技术，与传统的照明手段相比，这种技术可以将能耗降低10倍、12倍甚至15倍。且每年灯光节结束后，主办方还非常注意布景材料的回收循环利用。

2011年里昂灯光节

2011年，里昂灯光节融合当代科技和设计艺术，打造一场听觉和视觉的灯光盛宴。该届灯光节有"云端彩梦"、"生生不息"、"城市百态"等不同主题。

"云端彩梦"由Jacques Rival设计，一束巨大的彩色发光气球漂浮在白苹果广场近两个世纪的路易十四世骑马雕像上方，悬挂在天与地之间，25米的高度承载着彩色的梦，像要冲破天际飞到云端去一样，就连孤独了近两百年的路易十四雕像，在彩色气球的带领下，仿佛也走上了新的旅程。

"生生不息"是新加坡一流艺术家Sun Yu-Li和Allan Lim先生带领团队，为2010年的Light Marina Bay而设计的。它周身炫目而平和的蓝色光芒像是用一种光的语言向参观者传递着保护环境、节约能源的信息。这个高8米的作品从2011年8月开始制作，采用不锈钢结构，由1 000个废弃塑料瓶和3 000个星巴克的废弃塑料杯做成的塑料花组合而成。灯具选用方面也依据环保节能和绿色生活的理念选择了philips的color kinetics低功

率 LED 灯。

而"城市百态"是由设计师 Thomas Veyssiere/ Groupe LAPS 设计，66 个 LED 制作的电子人展示出人的各种运动、表达不一样的情绪和状态。一段东方音乐响起，电子人逐次亮灭，仿佛一个人在原地练武术，然后是两个人武术对练；一段悠扬的圆舞曲响起，电子人翩翩起舞，有独舞，也有男女共舞；音乐高亢，所有电子人都振臂高呼，以表达人的蓬勃朝气。整部音乐编排得激情四射，让人情不自禁地翩翩起舞起来。

2012 年里昂灯光节

里昂是著名的画壁之都，也是电影创始人卢米埃尔兄弟 (Lumiere Brothers) 的故乡。2012 年里昂灯光节以投影为核心亮点之一，给墙面"穿"上特别的感光涂料，通过光线作用达到透视的动感效果，可以在墙上作画、演戏、讲述历史。

里昂灯光节里的另一大亮点是灯光作品具有强烈的互动性。它们将灯光的概念转化成以灯光和色彩为基础的互动系统，提高人和光之间的互动性，使观众更深入地参与到节日中。

1. 神奇的立方盒子

在 2012 年里昂灯光节上，灯光设计师 Gilbert Moity 展出了他所创作的独一无二的交互式和试验性的灯光作品——神奇的立方盒子。这个作品起源于 2003 年，当时 Gilbert Moity 通过一个建筑项目发现了一种名为 Danpalon 的聚碳酸酯新材料。在对其进行光衍射的几次实验中，取得了令人惊讶的结果，这让设计师萌发了将光与 Danpalon 结合起来的想法。从那以后，投影仪表现出了足够好的性能，使制造大型动态光面成为可能，并催生了一个 360 度光面的立方光体。

建立一个光体的基本原理很简单：通过传感器展示不同的图像即可。但是设计者期望突破理念与技术的限制做出更特别的效果。由此促使了这个装配了发电机的自行车装置的产生，自行车产生的电能使光体产生各种色彩斑斓的创像。

为了让这个立方体变得更加吸引人、更加壮观，同时赋予其故事，Gilbert Moity 创造了一个用 28 辆自行车环绕青铜立方雕塑体四周的造型设计，让参与者沉醉于不断变幻的视觉画面里，营造一个兴奋、狂热的氛围。光体随着参观者踩动自行车而开始运动，速度随着图片逐渐加速变化而逐渐增强，最后以雕塑四周被绚丽灯光和烟花所围绕画上句号。光体长 8 米，宽 6 米，高 5 米，烟花高度达到 7.5 米。

2. 装置雕塑

灯光雕塑是近年出现的一种新型艺术表现形式，每年的里昂灯光节都能看到它身影。它将动态的光元素与静态雕塑设计融合，让古老的雕塑艺术焕发出新的气息，也点亮了现代城市的夜景。

3. 虚拟的大鸟笼

12 个柔软轻盈的巨型发光气泡将被放置在 Grande Cote 花园里。冬天的夜晚里，虚拟的鸟儿们会五个为一组，一起寻找庇护，被这些温暖的灯光紧紧包裹着。每个鸟巢就好比皮影戏，让人想起那些总是不断迁徙的大鸟，如同被风带进生活中的移动物品。伴随着歌声的引导，行人漫步在这些抽象的作品当中，与此同时，无论是观众还是表演的人，都将在这样的场景中唤起儿童时代的情感。

2013 年里昂灯光节

2013 年里昂灯光艺术节共展出世界各地的 120 位艺术家的灯光装置艺术作品。灯光师、音响师、数码图像师和造型艺术师用灯光、音乐、烟花在里昂市中心和周边 80 余处地点打造了风格主题各异的灯光秀。激光投影、3D 特效等技术被广泛应用于一个个灯光作品中。

沃土广场的"灯光王子"在两面墙上演绎，以"被盗的光明"为线索，讲述了居住在明亮星球上的小王子，历经艰辛，寻回光明的故事。整场灯光秀梦幻、童真，向人们展示了一个现代光学技术创造出的魔幻世界。位于里昂市中心的白勒库尔广场则展示了巨型摩天轮灯光秀。在布尔斯小广场，设计师利用投影"变"出了一束束灿烂的繁花。里昂市政厅的外墙上，也演绎着灯光秀"消失的天堂"。

此外，2013 年里昂灯光艺术节第一次设立"中国角"。"中国角"项目由中国留法艺术家设计并策划，在里昂金头公园竖起了一面红灿灿的灯笼墙，由 1 350 个红色灯笼组成。这些灯笼的原材料大部分从广州运来，在里昂当地由灯光艺术节工作人员组装完成，成为两座城市合作而成的艺术品。

018 | 流光溢彩 灯光设计
THE COLORS LIGHTING DESIGN

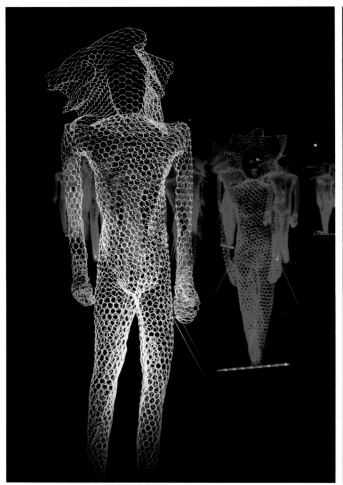

Lighting Art Festival, actually a regional festival for Lyon, lasts for 4 days from December 8, during which each household displays burning candles outside window to create candela shining and glittering throughout the whole city. And so do religious avenues and some other public places, including Basilica of Notre Dame de Fourviere and Place des Terreaux in particular.

It's recorded in the history that there is a legend behind Lyon Lighting Festival, that in 1852, the Cathedra in Lyon was rebuilt with the over-gild copper status of Mother Mary. Celebrations had been planned on September 8, birthday of Mother Mary. As unexpected, commemorative activity was postponed until December 8 because of the soaring water in the river of Sohne. On that morning because of a heavy rain, when the archbishop would claim that celebrations had to be delayed for another day, local people becoming frustrated and disappointed for it, the raindrop came to a halt between the lights, the sky cleared up. Exhilarated people lit up candles by the window, and then poured onto the street dancing and singing. Light from candle died out until the next dawn. The day of December 8 consequently started the prelude to hold Festival of Lights.

The 2011 Lyon Lighting Festival

In 2011, Lyon Lighting Festival turned out to be a visual-audio lighting feast with modern science and design combined. A grand event it made with numerous themes, like "Dream above Clouds", "Vigor and Vitality" and so on.

"Dream above Clouds" by Jacques Rival, consists of a huge bunch of colorful shining balloons flying above the status of Louis XIV who almost rides on a horse there for two centuries. Between the sky and the land, the height of 25 meters with color dreams seems to have been break through the skyline and beyond clouds, like the lonely Louis who has been on a new way led with the balloon.

The team lead by Sun Yu-Li and Allan Lim designed "Vigor and Vitality" was for Light Marina Bay 2010. Stunning yet peaceful, the blue throughout the whole body conveys a message with lighting vocabulary to protect environment and save energy. The project measuring 8 meters high was of stainless steel structure. The process with 1,000 plastic bottles and over 3,000 plastic cups collected in Star Bar started from August, 2011. LED lighting of Philips and Color Kinetics are used for their advocating environment-friendly and green concept.

As for "City Epitomes", the cyborg of 66 LED pines by Thomas Veyssiere, Groupe LAPS exhibits all kinds of human gestures, emotions and confronting status. Pipes come on and off to the music, like a man staying where he is, practicing China's martial art; and then two men are fighting. When waltz begin, some cyborgs perform sole dance, some dance with others. When music reaches its peak, all cyborgs are raising their arms and shout to express human vitality. Music throughout is filled of zeal and passion, people naturally singing and dancing in a happy mood.

The 2012 Lyon Lighting Festival

Capital of mural, Lyon is the birthplace of Lumiere Brothers, founder of film. Projection was taken as another light spot for the Lighting Festival in 2012, by which walls are coated in special light paint. Ray of light makes perspective effects on the wall either by picture drawing, play acting or storytelling.

Lighting works are endowed with strong interaction. The concept of lighting is transformed into an interaction system based on lighting and color to promote the interactivity between views and light, so that people can participate in the event to a deeper extent.

1. The Magic Cubic Box

Lyon in 2012 saw the Magic Cubic Box, an interactive and experimental work by Gilbert Moity that dated back to 2003, when Gilbert was exposed to a new Makrolon material named Danpalon. The first experiments conducted by Moity led to findings, inspiring the combination of lighting and the materials. Projector has been exerted much better since then, its potential of large-scale dynamic glaze made possible. Meanwhile, to build up a cubic light object with 360-degree glace appeared vividly.

Its basic principle is easily understood: sensors are used to display different images. However, the limitation of theory and technology is aimed to be broken through to generate special effect. The final result is the device looking very similar to bicycle but equipped with electric generator, whose power shifts light energy into colorful images. In order to make the cubic more attractive and notable, and allow for its connotation, Gilbert brought forward a cubic bronze sculpture surrounded with 28 bicycles. When the participants are lost in the changeable visions, an atmosphere is made excited and zealous. When views ride on the bike, the object sets its motion. Speeding picked up with more pictures, the sculpture is embraced with lighting and firework. The light body measures 8 meters long, 6 meters wide and 5 meters high, and the firework can reach as high as 7.5 meters.

2. Installation Sculpture

As a recent art form, lighting sculpture is available in Lyon every year, which fuses the dynamic light element and the static sculpture to refresh the old art in lighting up the nocturne in modern cities.

3. An Imaginary Aviary

12 huge bubbles are put soft and light into the Garden of Grande Cote. On winter nights, birds imaginary are clustered in warm light, five per group. Well reminiscent of those constantly in migrant states, each bird's nest seems to be implanted into a shadow play like mobile items forced into people's sight by wind. With the guidance of music, pedestrians meander through the imaginary objects. Emotions in childhood are evoked in the depth of people's heart of both views and performers.

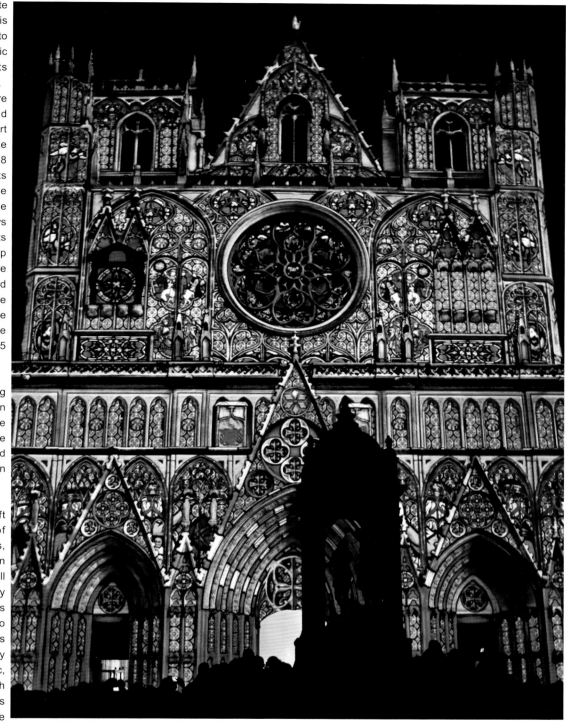

The 2013 Lyon Lighting Festival

Lamplight installation were done by 120 artists across the world in 2013. Beneath the finesse, ingenuity and skill of engineers of lighting, sounding, digital image and modeling, over 80 spots in Lyon and in its neighboring communities presented varying styles and themes with lighting, music and firework. Laser projection and 3D special effect are applied to shape light works.

Lamplight Prince on Place des Terreaux is interpreted on both sides: the stolen light is taken as the clue to tell a story, that the little Price from Star of Brightness regained brightness after going through thick and thin, making a magic world with modern optical technology. In the Kurt Square in the city stand a large sky wheel to show lighting. In another smaller square, projectors are used to accomplish bunches of flower blossoms. And the façade of the city hall performs another lighting show of Paradise Lost.

Additionally, Chinese Corner was established for the first time in 2013, a project designed and planned by artists from China. A wall stands red, of 1,350 of lanterns. Most of materials involved are imported from Guangzhou, and manufactured on the spot by staff working for the Lighting Festival. A real art work it makes with cooperation by two cities.

流光溢彩 灯光设计
THE COLORS LIGHTING DESIGN

"神秘博士"别墅
THE HOUSE OF DOCTOR WHO

《神秘博士》由BBC制作,并于1963年11月23日首次在英国播出,这部渗透了英国文化的科幻剧集在英国电视迷中的地位犹如《星际迷航》在美国人心中的地位一样重要。不仅在英国广受欢迎,《神秘博士》的吸引力也波及全球,几十年来魅力不减。

在英国广播公司BBC推出这部经典剧集50周年之际,Spinifex Group利用变幻的光影、扣人心弦的音乐、逼真的视觉效果,在澳大利亚悉尼环形码头的海关大楼打造一场炫目的3D灯光盛宴。海关大楼变成了一座露天剧场,大楼的外墙被用作天然大银幕,向游客展现了"神秘博士"在时空中穿梭、与黑暗势力抗衡的扣人心弦的剧情。

Spinifex Group希望以这种特殊的方式向这部荧屏经典致敬,同时,这也是2013年"活力悉尼"灯光音乐节的一部分。

Doctor Who, is a sci-fi series by BBC broadcast on Nov. 23, 1963, and has really been occupying a vital role for British TV show fans, quite like Star Trek for American. Additionally, its attraction has been overspread around the world and has witnessed any reduction for decade.

On its 50th anniversary, Spinifex Group presented a stunning 3D-mapped projection tribute to the science-fiction series, which was screened on the grand facade of Customs House at Circular Quay by optimizing the chord of light and shadow, live music and vivid visual effect. An outdoor theater the Customs House made, where The Doctor tackles his foes, shuttling back and forth in the space and countering against the evil in a gripping story.

A special approach by Spinifex Group is destined to pay homage to the classical screen, when providing a part for Australia's Vivid Sydney Festival.

室外灯光
OUTDOOR LIGHTING 023

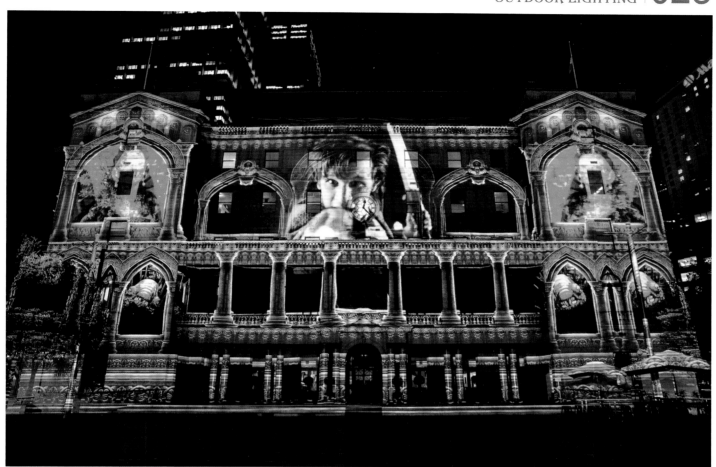

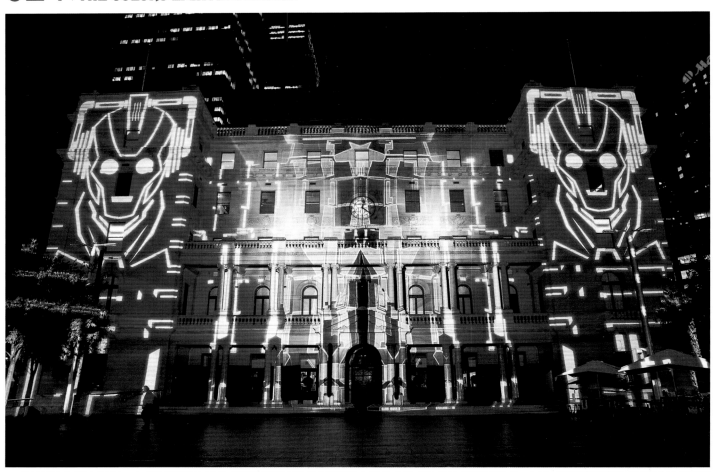
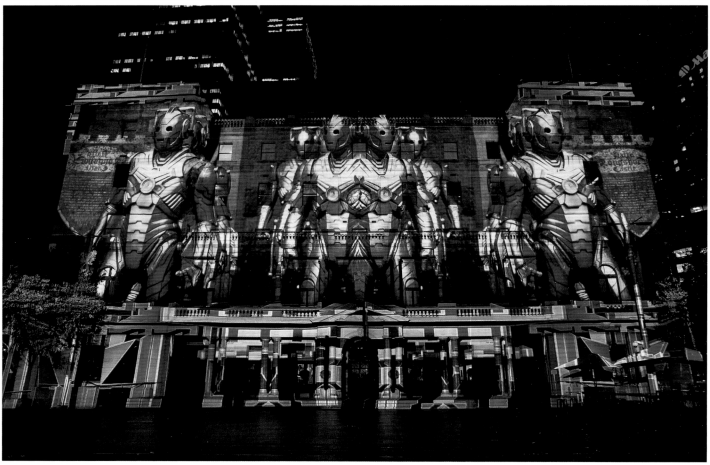

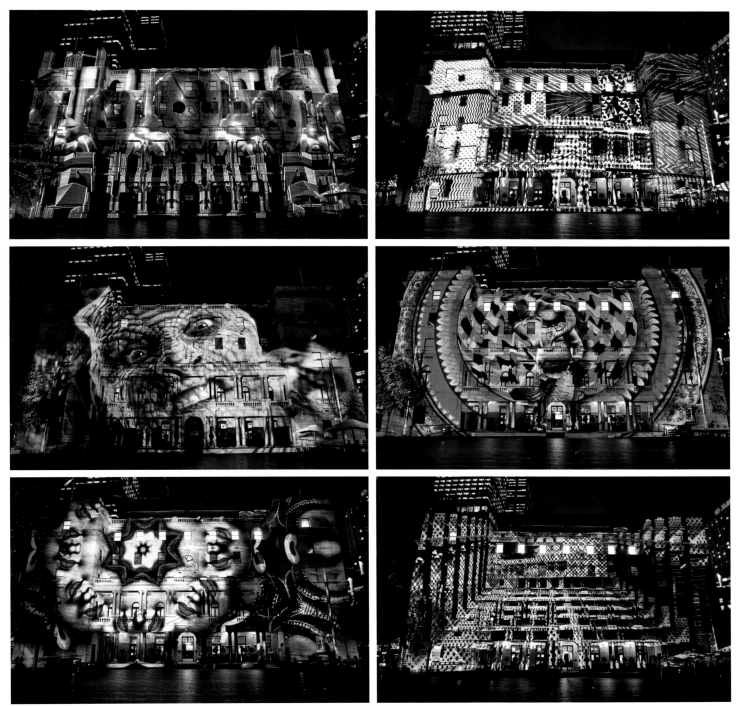

悉尼歌剧院灯光展
THE SYDNEY OPERA HOUSE LIGHTING EXHIBITION

流光溢彩 灯光设计
THE COLORS LIGHTING DESIGN

设计公司：三齿稃集团、59制作机构　　Design Company: Spinifex Group, 59 Productions

"活力悉尼"灯光艺术节是集灯光、音乐和创意于一体的年度艺术盛会，这场视听盛宴每年在 5 月底至 6 月中期间闪耀登场，用创意和灵感为城市画上缤纷的色彩，并点亮万众期待的悉尼歌剧院白帆灯光，向世界展示悉尼作为亚太地区重要创意艺术中心的风采。

2013 年悉尼歌剧院灯光展以"玩味（PLAY）"冠名，拉开了年度灯光展的狂潮。轻启按钮，15 分钟的灯展即刻拉开帷幕。诸种风格，万盏华光，随即从一个舞台辗转至另一舞台。场景有生机，有优雅，动感、大胆、愉悦地玩味于艺术与建筑之间，触动着悉尼人的心灵，书写着"生动"的含义。

此次悉尼歌剧院灯光展设计希望能呈现一种超越于艺术的动机，实现一种颠覆的现实。于其而言，"玩味"的焦点首先在于娱乐。设计师希望本案在生动讲述故事的同时，收获世人的尊重，激发起观众的思想。每个人的心目中都有取舍，获其至爱。人如此，悉尼也应如此。悉尼有多样、生动的创意文化。能于悉尼看到一个灯光悉尼，是人们爱上悉尼的另一原因。这是一种玩味。而工作与设计本身，也是一种玩味。

2014 年悉尼灯光音乐节，悉尼歌剧院再次盛装出现于江湖。从概念到设计，到建筑，至内里的各种饰面，色调及质感，无不尽其精彩。设计以展现空间未来形象为主旨。在经历从自然极致因素至近乎毁灭的境地，至宇宙神游，再至最后：当一切归于平静，天地之间依然是一种成功的典范。

空间希望能引导观众进入一种旋风般的旅程，通过从建筑至竣工的重新诠释，探索歌剧院的概念与设计，借助于强烈的舞蹈展示及灯光的序列游动，感受质感、色度、服装甚至发型，人们可以感受到建筑的宏伟。明亮的色调、墨水及初始状态的颜料款款地出现于世人面前。

其中，灾难性的暴风雨、极致天气会给建筑带来终极考验。而最终，当一切风平浪静，又展示给人们一个万物重生的清朗世界。

室外灯光
OUTDOOR LIGHTING

030 流光溢彩 灯光设计
THE COLORS LIGHTING DESIGN

The 2013 Sydney Opera House Sails projection, spearheaded this year's vivid light festival, appropriately titled "PLAY". The 15 minutes show begins with a Play button, that launches into a restless journey through various style and genre's as it moves seamlessly from one scene to the next, exploring a distinctive mix of iconic and striking scenes, blended with vibrant and graceful movement. This energetic, daring, and delightful play between art and architecture touched the hearts of Sydney Siders and reflected what Vivid is all about.

Spinifex Group intended the piece to move beyond just art, and become a head turning show. In order to title the piece "PLAY" the focus was on entertaining first. Richard Lindsay, Head of Creative, "We'd love to earn the respect of the audience and give Vivid a story to tell. Our evolving journey approach will hopefully evoke the audience's thoughts and opinions. Perhaps they will have a favourite section; perhaps Sydney will have a favorite section? We look forward to the public reaction. Sydney has a diverse and vibrant creative culture, but it's important to see the lighter side of life—that's why we love Sydney. This is a fun piece, and the Spinifex team had a lot of fun creating it".

This year, you will see the Opera House born again, from its conception and design through the extraordinary construction process, before experimenting with an extraordinarily diverse range of surface finishes, colors and textures. Projecting the building into an imaginary future, it is then subjected to extremes of natural forces and reaches a point of near-destruction, before undergoing a triumphant apotheosis, and beginning its cosmic journey once again.

The artwork of 2014 whisked the audience on a whirlwind journey that explores the conception and design of the Opera House, through a recreation of the spectacular construction process to completion. The magnificent building can then be seen expressing itself in a high-octane dance and light sequence, experimenting with texture, color, costume and even its own haircut, before being doused in brightly colored paints and inks, and showered in raw pigment.

In the third act, a cataclysmic storm and extremes of weather and climate put the building to its greatest test, before reaching its apotheosis, atomizing into cosmic form before reforming and beginning its cycle once again.

流光溢彩 灯光设计
THE COLORS LIGHTING DESIGN

中国·哈尔滨国际冰雪节
HARBIN INTERNATIONAL SNOW AND ICE FESTIVAL

"中国·哈尔滨国际冰雪节"不仅是中国历史上第一个以冰雪活动为内容的区域性节日，同时也与日本札幌雪节、加拿大魁北克冬季狂欢节和挪威奥斯陆滑雪节并成世界四大冰雪节。

而作为冰雪节三大重要冰雪景观之一的哈尔滨冰灯艺术游园会自1963年创办迄今已成功举办了41届。中国冰灯艺术发源于兆麟公园，是以冰为载体，以灯为灵魂，融绘画、雕塑、建筑、园林、文学、音乐等多种艺术于一身，并可运用光学、力学、声学及电气、机械等科学技术的新兴的造型艺术。运用这种艺术形式创造出的珠宫琳馆、琼榭瑶桥、银雕玉塑、冰山雪岭、冰凌冰瀑，白昼间千姿百态、晶莹剔透，入夜后流光溢彩，万紫千红，像扑朔迷离的仙山琼阁，似如梦如幻的水晶宫阙，且年年出新意，岁岁景不同，被称为"永不重复的童话世界"。

哈尔滨冰灯艺术游园会经过多年举办，在组织策划、设计建设和内容方面不断推陈出新，令人瞩目。

2010年，第37届冰灯艺术游园会汲取历届之精华，以"再现冰灯艺术经典·演绎冰雪多彩华章"为主题，充分利用兆麟公园地势地貌环境特点巧妙布局，规划设计了"盛世凯旋"、"历史博览"、"卡通乐园"、"冰上圆舞"、"银龙盘舞"、"流光溢彩"、"世博之光"、"冰情雪韵"、"秀美江南"、"冰雪竞技"十大景区，用冰雪量3万立方米。园内以冰雪艺术为主导，彩灯、灯雕、彩冰、冰花、脸谱与LED灯效等多种艺术形式巧妙结合，相生相谐。

2011年，第38届哈尔滨冰灯艺术游园会以"冰雪誉华夏·精彩传世界"为主题，总用冰雪量预计达2万立方米，景区规划设计凸显"中国风格"，依托兆麟公园自然环境和地势地貌规划，设计"开篇盛世"、"秀美江南"、"荷塘月色"、"冰雪之源"、"盛世欢歌"、"儒思千年"、"冰艺竞技"、"璀璨奇葩"八大景区。景区利用低碳环保太阳能板，通过白天吸收太阳光储存转化成电能，与LED灯具结合应用，打造绿色环保冰灯景观。此外，冰灯游园会还采用彩冰、彩灯、冰花、灯雕、LED七彩变光以及非冰环保制品等多种艺术表现形式与冰雪园林有机结合，独特艺术冰版画和彩冰书法也穿插其中。

2012年，第39届冰灯艺术游园会按照"传承、发展、丰富、创新"的原则，以"浓浓冰雪情·印象哈尔滨"为主题，以哈尔滨意蕴深厚的历史文化为创作蓝本，依托兆麟公园自然环境和地势地貌，规划建设"东方印象"、"江畔记忆"、"西域丽影"、"跨越时代"、"冰艺奇葩"、"情定滨疆"、"青艺技竞"、"童乐趣谷"八个景区，荟萃冰雪艺术作品2 000余件；聚集了冰城百年胜景和风情灵秀，结合多种艺术形式将哈尔滨人文历史和城市风情寄寓在冰雪景观创作中；用冰雪故事娓娓道来，唤醒历史记忆。同时展示了哈尔滨"音乐之城"、"冰雪之都"的城市形象，使游人仿佛漫步于浪漫迷人的"东方小巴黎"。

2013年，第40届哈尔滨冰灯艺术游园会以"冰灯五十年，精彩传世界"为主题，依托兆麟公园地势地貌规划设计"奔腾"、"冰艺荟萃"、"游乐天地"、"欧陆风情"、"北疆冬趣"、"冰雪之源"、"盛世之光"、"锦绣中华"八个景区。主景楼设计为欧式风格，高5米，与中央大街相呼应。楼上建观赏区，游客可登上去看遍园区内五彩缤纷的冰灯景观，也可面朝中央大街，欣赏冰城的欧陆风情。西门打造三叠层的哈尔滨元素印象派城堡，

室外灯光
OUTDOOR LIGHTING | 035

036 流光溢彩 灯光设计
THE COLORS LIGHTING DESIGN

火焰和古堡造型为巴洛克和哥特式建筑风格，彰显哈尔滨城市特色，远看波澜壮阔，近观微妙无穷。

When making the debut themed with ice and snow activities in China's history, Harbin International Snow and Ice Festival has been ranked top four with Sapporo Snow and Ice Cream, Quebc Winter Carnival and Oslo Skiing Festival .

As one the Three Top attractions during Harbin Snow and Ice Festival, Harbin Ice Lantern Festival has celebrated its 41st anniversary since its beginning in 1963. Originating from Zhaolin Park, the Garden Party employs ice as its carrier and soul as its soul in integrating into one

painting, sculpture, construction, gardening, literature and music. Meanwhile, multiplied forms of science and technology are used, covering optics, mechanics acoustics, electric and machinery. The result of peal palace, jade shed and bridge, silver carving, iceberg, snow ridge, and ice cascade, is throbbing as clear as crystal during daylight and equally shining at night. The immortal mountains and the pavilion of fine jade with the palace one after another is rebirthed year after year, but never making replicas with any similarity to the previous one, which wins it a title of A Never Repetitive World in Fairy Land.

The attention it captures is owed to its organization, plan, design, construction and contents destitute of the old state but completely-fresh. The year of 2010 witnessed the occurrence of the 37th, one that withdrew the essential quality from those before. In order to highlight the theme of "Recurrence of Classical Ice Lamp and Interpretation of Charming Ice and Snow", the geographical feature of Zhaolin Park is fully optimized to accommodate 10 Sceneries, like "The Time of Prosperity", "The Expo of History", and "The Paradise of Cartoon" and so on. The ice and snow used reached 30,000 cubic meters. Ice and Snow art dominated the park then, when illumination, ice carving, stained ice, ice mask and LED effect were fused into accord and harmony.

In 2011, the 38th Harbin Ice Lantern Festival referred to "The Ice-Snow Fame in China and The Ice-Snow Marvel into the World" as its theme, with 20,000 cubic meters ice and snow involved. Also based on the physical conditions of the park, the event then sets off "China's Style" with 8 Sceneries, like "The Prelude of Prosperity", "The Beautiful Low Reaches of Yangtze", "The Moonlight over the Lotus Pond" and so on. Compared to those former ones, the one that year first turned to PV panel to take in solar energy to generate power, which with LED lighting made green and low-carbon scenes. Apart from the familiar elements like illumination, lamp carving, 7-colored LED fitting and items not of ice organically with the gardening wrapped in ice and snow, ice

038 | 流光溢彩 灯光设计
THE COLORS LIGHTING DESIGN

OUTDOOR LIGHTING 039

ocks and works of lligraphy of colored e are interspersed and erwoven.

icking to the principles f "Inherit", "Develop", nrich", and "Innovate", e 39th took as its eme "Affection from e and Snow, and arbin Impression" in 12, when the profound cal culture was made the blue print, and eanwhile, 8 scene ots were housed sed on features of the rk geography, like "The pression of the East", he Memory along the ver", "The Images of e West Land", etc. ver 2000 ice-snow art eces were available. estinations grand d great, and scenery quisitely beautiful re embedded into the snow landscape while ories about ice and ow were told vividly, miniscent of the past. the 39th event, the ty image of Harbin "city of music", and apital of ice and snow" re displayed, and rticipants seemed to meandering in the ster Paris.

for the theme of e 40th in 2013, it's e Lamps through 50 ars, and "Marvel into e World". Another 8 enic spots were made the basis of the park' terrain and landform, e "Gallop", "Collection Ice Art", "Land of tertainment", "Charm ross Europe", "Light Prosperity", etc. The ain building was as gh as 5 meters and of ntinental style, echoing th the Center Avenue. he part upstairs was

040 | 流光溢彩 灯光设计
THE COLORS LIGHTING DESIGN

especially reserved for sightseeing, where visitors could have a panoramic view, including the Center Avenue and the European charm of the ice city. Around the west gate was an impressionism castle, a world of Harbin elements. Like the firework, the castle is Baroque and Gothic to manifest the city's features: sights come into view if looked far away while subtlety is perceivable when watched nearby.

悉尼海港大桥灯光秀
SYDNEY HARBOR BRIDGE LIGHTING SHOW

设计公司：齐合作设计　　　　　　　　　　　　　　　　Design Company: Chiho & Partners

悉尼海港大桥是早期悉尼的代表建筑，它像一道横贯海湾的长虹，巍峨俊秀、气势磅礴，与举世闻名的悉尼歌剧院隔海相望，成为悉尼的象征。

作为世界闻名的地标建筑之一，悉尼海港大桥在2013年活力悉尼灯光音乐节举办期间，被装饰成为世界一流的灯光艺术雕塑的中心焦点。大桥的灯光艺术设计由活力悉尼活动伙伴澳洲英特尔公司（Intel Australia）携手悉尼灯光创意公司32 Hundred Lighting共同创作，通过互动及可程控的灯光装置，将身临其境的光影艺术投射到大桥的西面，使公众可首次透过互动触控屏幕控制灯光投射。32 Hundred Lighting负责包括100 800组独立并可受控的LED灯饰和划时代互动触控屏幕接口软件程序的设计。

As one of the early building representatives, opposite to Sydney Opera across the bay, is like a rainbow that flies lofty and pretty with great momentum in the sky.

During the period of the 2013 Sydney Lighting Festival, the world-renowned Australian landmark is decorated as the flagship attraction. The installation of the bridge is done with cooperation of three partners, including Intel Australia and 32 Hundred Lighting. The lighting device of interaction and producer control projects light and shadow onto the west of the bridge, and the public then can regulate the lighting projection by touching the screen. The LED lighting of 100,800 pieces and the software for the interactive screen are under responsibility of 32 Hundred Lighting.

室外灯光
OUTDOOR LIGHTING | 043

流光溢彩 灯光设计
THE COLORS LIGHTING DESIGN

巴库水晶大厅
BAKU CRYSTAL HALL

设计公司：GMP 建筑师事务所
灯光设计：Lichtvision 设计

Design Company: GMP Architekten
Lighting Design: Lichtvision Design & Engineering GmbH

"巴库水晶宫"的设计其实是为 2012 年欧洲赛歌会提供一个聚会的地方。现代多功能的空间可以容纳 23 000 名观众。工程迅速，仅仅耗时 8 个月。206 米长，168 米宽的钢结构覆盖上一层华彩的拉膜，彰显建筑水晶透明的特点。夜晚时分，主厅立面由 5 400 个 LED 光源照耀，视觉倍感宏伟、浩大。

Just in time for the Eurovision Song Contest 2012, the Baku Crystal Hall–a modern, multi-functional event arena with a capacity of 23,000 spectators-could be completed in a record time of only eight months. The 206-metre-long and 168-metre-wide steel construction is clad with an illuminated membrane facade, which lends the building its characteristic crystalline appearance. During the dark hours the majestic hall's facade is lit by 5,400 LED RGB light points that make a great visual impact.

室外灯光 OUTDOOR LIGHTING 045

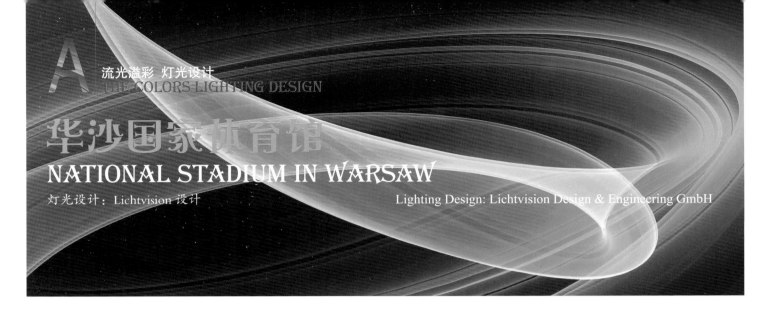

华沙国家体育馆
NATIONAL STADIUM IN WARSAW

灯光设计：Lichtvision 设计　　　　　　　　　　　Lighting Design: Lichtvision Design & Engineering GmbH

华沙国家体育馆地址近乎城市中心。设计的目的是在华沙树立一个新的地标式建筑。"建筑的整个立面如同静止的地标，或者当做低分辨率的屏幕使用"是整个设计的中心思想。不同寻常的灯光转变着建筑立面使用材质的质感，强调着立面的节奏。夜晚，立面的透明金属网如同遁形。外观的立面映照着五个动感光舞场景，如"星空"等。另外，还有三个静态光感现场于立面上彰显。

The stadium in Warsaw is located near the city center. The design team's main brief was to transform the new landmark building at the centre of Warsaw at night time. The main idea was to use the whole facade as a static landmark or a low resolution screen. The illumination transforms the facade material, so the lighting effect is extraordinary, emphasizes the rhythm of the facade architecture and closes visually the translucent metal mesh facade panels at night. Based on pre-programmed scenes, moods from within the stadium can be projected onto the facade. Five dynamic lighting scenes (e.g. starry sky, La Ola wave, scores, dynamic encircling lettering) and three static ones have been pre-programmed for the exterior facade.

室外灯光 | OUTDOOR LIGHTING | 047

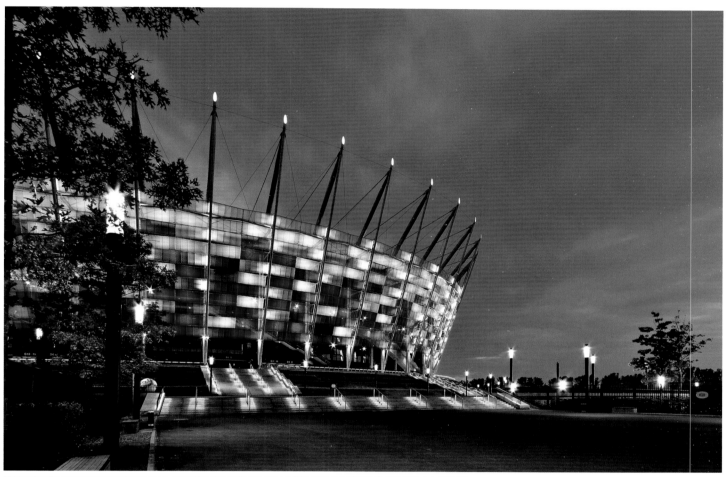

048 流光溢彩 灯光设计
THE COLORS LIGHTING DESIGN

流光溢彩 灯光设计
迪拜亚特兰蒂斯酒店开幕灯光秀
THE PALM GRAND OPENING LIGHTS SHOW, ATLANTIS

照明

本次照明展由5.1环绕立体色，32台放映机向外界转播。66万个各式照明，相当于500个100瓦的灯泡参与了盛世灯展。

烟花

"烟花"由格鲁奇设计执导。贝壳状的造型升腾至274米的高空中，远远可见。整个工程消耗绳索长达79千米。75名员工组成的团队，历时14天把其伸展至520千米之外的朱美拉酒店。差不多一半的烟花造型利用的烟花火源达到了716个。爆炸时的棕榈树造型全为定制。空中最大的贝壳造型可升腾至高空300米的地方，爆炸跨度达200米，相当于两个足球场的面积。夜晚的礼花着红、着白、着绿，照亮着亚特兰蒂斯及朱美拉大酒店。来看中国、西班牙、意大利、美国的烟花是对酒店客人的礼赞。10万个特别设计的烟花装饰表演期间，不到9分钟的时间，全部点燃、绽放。

The Illumination

The visual illumination show was broadcast in 5.1 Surround Sound and transmitted by 32 projectors, producing 660,000 lumens, the equivalent of almost 500 100-watt light bulbs.

The Fireworks

The fireworks display designed and executed by Grucci, featured some shells bursting at an altitude of 274 meters and visible from space. The pyrotechnics production utilized 79 kilometers of cable, which had been stretched across the 520 kilometers of Palm Jumeirah by a team of 75 people, over14 days' prior to the event. Half of the fireworks utilized from 716 firing locations were custom-made palm tree-shaped bursts. The largest aerial shell rose 300 meters before its flowery burst spanned more than 200 meters in length and was equivalent in size to two football fields. In true Kerzner tradition, the lost city of Atlantis arose from the Arabian Gulf in a water and light spectacle. The night's pyrotechnics, which also illuminated Atlantis and Palm Jumeirah in red, white and green in homage to Atlantis' newest home, the United Arab Emirates, were shipped in from China, Spain, Italy, and the United States. More than 100,000 specially-designed pyrotechnic devices were fired in less than nine minutes during the multi-luminous pageant.

室外灯光 OUTDOOR LIGHTING | 051

筒仓 468
SILO 468

流光溢彩 灯光设计
THE COLORS LIGHTING DESIGN

设计公司：灯光设计集公司　　　　　　　　　　Design Company: Lighting Design Collective

借助于设计妙手，原本已经废弃的仓库焕然发出了生命的活力。华丽转身间，一个新颖的艺术装置成就了一个崭新的公共空间。"筒仓468"，废弃改建项目的一个工程开启了本案所在滨水前沿的重新发展，为世人呈现了一个地标式的建筑景观，几公里之外却可以清晰可见。

自然的、人工的用材为人们提供了一个重要的去处。定制的软件，1 280 个 LED 灯组装饰的穹顶顺应着风速、风向、气温而做出适时的变化。或是流畅，或是自然，或是与众不同，如同一个熠熠生辉的壁画，美轮美奂。

白天的日光顺着开口的洞，洒下一路的光芒，内里一片深红。450 个钢构镜面置于洞后，反射着太阳光线，临近的水面波光无限。

Madrid-based lighting design collective have converted a disused oil silo into an art installation and new public space. "Silo 468" signifies the start of a major waterfront redevelopment in helsinki, and is intended to create a landmark for the district, visible from several kilometers away.

Lighting design collective have made use of both natural and artificial light in developing the scheme which forms an important civic destination. Through the implementation of a bespoke software system, 1,280 LED domes are able to react to the environmental conditions, changing in relation to the wind speed and direction, temperature and weather. The occurring patterns are designed to be fluid, natural and unique, and the result is a constantly evolving glowing mural.

The interior of the building is painted a deep red, as daylight seeps in through the perforated walls. 450 steel mirrors placed behind the holes reflect sunlight, allowing the project to sparkle mirroring the surface of the adjacent water.

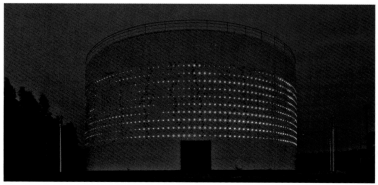

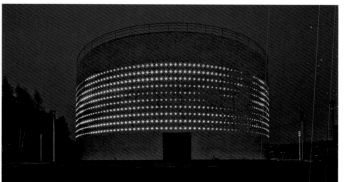

室外灯光 | 053
OUTDOOR LIGHTING

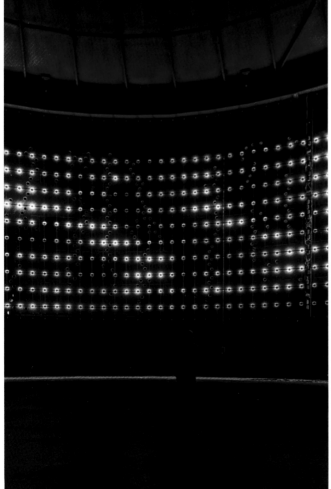

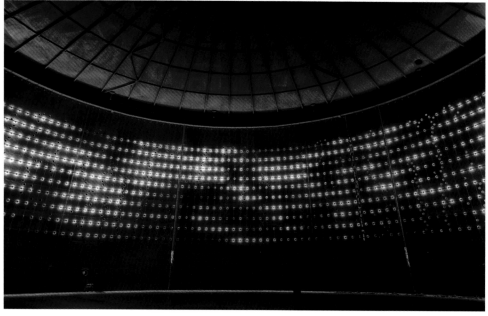

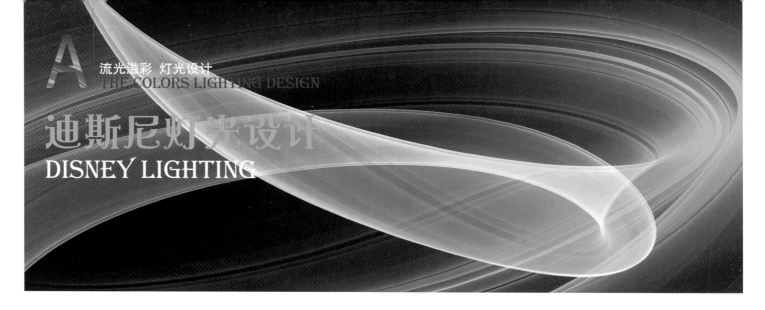

流光溢彩 灯光设计
THE COLORS LIGHTING DESIGN

迪斯尼灯光设计
DISNEY LIGHTING

对于本案而言，设计过程源于概念。设计首先要搞清具体空间需要什么样的灯光。需要保留无光照的黑暗状态的空间因刻意"留黑"。"留黑"空间的管理、设计如同光照本身一样重要。这些问题一旦搞定，便要考虑灯光以什么样的形式出现。来光的角度、光束的宽度、光源的置放等都要予以考虑。

光亮时，生命便会点燃。色调便是光的生命。即便白光，也有种种，有的呈蓝色，有的呈橘黄色。有的白光配色很好，有的白光会使其他色彩相形见绌。这便是光的渲染。

因为光照有其自己的生命临界，本案设计所用灯光尽可能维护简单，高能而且符合光照的相关规定。为了便于维护，所用照明全为 LED。当日光一点点消失，黄昏渐变成黑暗。华灯初上的瞬间，观众会为之完全惊叹。夜间的魔术就要上演。

为了方便照明的操作管理，"迪斯尼乐园"使用的全是电脑程序化管理灯组。该程序可用年限长，达 20 年之久。虽说简单，但却值得依赖。而沃特迪斯尼工作室公园，使用的灯控系统更为高端，操作性能更为精准。通过该系统，户外的照明可实现自由开关。

室外灯光
OUTDOOR LIGHTING 057

058 流光溢彩 灯光设计
THE COLORS LIGHTING DESIGN

The process starts in concept. Based on the story, the lighting designer will figure out what a particular room should look like. He or she will decide which parts of a particular space will have to be lit. Parts that may not be seen will deliberately be left in the dark. The lighting designer will also select the places where shadow should be dropped. Shadow management and shadow design is as important as lighting itself. Once it has been decided what needs to be lit, the lighting designer will determine how that should be done. He or she will chose the angle and width of the light beam and select the place where the light source should be installed.

The lighting designer will also have to pick the color of the light. When you put a light up, it has its own color. That's something people often tend to forget. There are all sorts of white light. Some is very blue, while other is very orange. The lighting designer will choose a type of white that works well with color. There are some kinds of white light that kill color, while others make the colors pop. This is called color rendition.

Things are here for ten, twenty, thirty, forty or fifty years. So the lighting equipment must hold up over time, despite heavy use and terrible weather conditions. It has to be as easy to maintain as possible. The lighting equipment must be energy efficient and comply with all applicable regulations. The maintenance team asked us to look at it again and to try and find lights that are easier to maintain. By that time, the moving light manufacturers had just gotten into the LED world. And using LED lighting seemed to cover our needs. When daylight is falling, twilight is deepening into night and it gets darker bit by bit. And then, all of a sudden, the lights come up and you see guests being totally amazed by it. It's when nighttime magic comes to life. So turning the outdoor lighting on is a huge thing!

In order to turn things on and off in an efficient way, we're using a computerized lighting control system. In the Disneyland Park, the system is 20 years old. It's a simple but reliable one. In the Walt Disney Studios Park, we have a more sophisticated lighting control system that enables you to do more precise things. The system allows you to program when and how the outdoor lighting should be switched on.

室外灯光
OUTDOOR LIGHTING | 059

B 室内灯光
INTERIOR LIGHTING

参考资料：光影良品、耐思空间

用灯光实现照明之外，还可营造美妙的空间氛围，调节使用者情绪，凭借光影变化丰富室内空间情境。

KTV灯光设计原则

一、功能性原则

灯光设计必须符合功能的要求，根据不同的空间、不同的场合、不同的对象选择不同的照明方式和灯具，并保证恰当的照度和亮度。即使是KTV，也未必需要太过花哨的设计。灯光首先要注意与房间大小协调，太过花哨的灯光色彩容易造成紊乱、繁杂的感觉，导致疲劳。每个建筑室内区域的灯光色彩必须协调搭配，体现层次感，分清主次，以达到美化空间的目的。例如大厅灯光可鲜亮明快，因为大厅是公共区域，所以需要烘托出一种友好、亲切的气氛，颜色要丰富、有层次、有意境，可以烘托出一种友好、亲切的气氛。

二、美观性原则

灯光是装饰美化室内环境和创造空间艺术气氛的重要元素。为了对室内空间进行装饰，增加空间层次，渲染环境气氛，采用装饰灯光，使用装饰灯具十分重要。设计师通过灯光的明暗、隐现、抑扬、强弱等有节奏的控制，充分发挥灯光的光辉和色彩的作用，采用透射、反射、折射等多种手段，创造温馨柔和、宁静幽雅、怡情浪漫、光辉灿烂、富丽堂皇、欢乐喜庆、节奏明快、神秘莫测、扑朔迷离等艺术情调气氛，为生活环境增添丰富多彩的情趣。

三、健康性原则

灯光色彩与人的心理健康有很大关联，例如蓝色可减缓心律、调节平衡，消除紧张情绪；米色、浅蓝、浅灰有利于安静休息和睡眠，易消除疲劳；红橙、黄色能使人兴奋，振奋精神；白色可使高血压患者血压降低，心平气和；红色则使人血压升高，呼吸加快，选择健康的灯光色彩有利于人的身心健康。

四、安全性原则

从建筑到室内、舞台以及空间的每一个角落，灯光的设计必须保证在安全性的原则下进行设计、安装、使用与维护。由于灯光来自电源，必须采取严格的防触电、防断路等安全措施，以避免意外事故的发生。

五、经济性原则

灯光并不一定以多为好，以强取胜，关键是科学合理的设计。灯光设计是为了满足人们视觉生理和审美心理的需要，使室内空间最大限度地体现实用价值和欣赏价值，并达到使用功能和审美功能的统一。华而不实的灯饰非但不能锦上添花，反而画蛇添足，同时造成能源浪费和经济上的损失，甚至还会造成光环境污染而有损身体的健康。从规划设计开始，经济性原则的考量就要贯彻始终，不但要保证建造之初的经济实用，还要在节能环保、日常使用、管理维护、更新升级方面实现经济平衡。

居住空间灯光设计原则

一、尽量保留室内的自然光线

评估室内承受的自然光线，如一天的光影移动，不同的天气情况下光影的明暗，选择舒适自然色彩，营造健康的灯光环境可尽量选用中性的淡色调墙面与地板颜色。此外，避免使用厚重的布料窗帘，不妨选用PVC材质（可遮挡反射太阳光的，雨伞用的布料）或罗马纱等透光性高的窗帘。此外，镜子的使用也是光线折射后让室内感觉明亮而舒服的布置技巧。

二、依需求选用不同灯具

根据不同的空间使用需求与使用者的心理需求，合理布局灯光，选择合适灯具。想要类似白昼的自然光线，可以考虑在天花板上悬挂吊灯、在地板安装立灯，或在桌上放置台灯；想特别强调空间的某个区域，可以选择水晶分枝吊灯，其光线折射效果可以表现多样化的戏剧性立体感。但对于某些机能性的空间而言，照明设计则显得特别重要。而感应式灯具更是比较贴心的照明设备，适用于玄关、走道、楼梯间或储藏室等小空间，节能照明。

三、不同光源，改变空间视野

只有光线具有方向性时，才会有光影的产生。如果想利用光线使居家视野更宽阔，可利用间接与反射灯光，让光线恍如淙淙流水，自然泼洒曳入；若想利用光线使居家视野更收敛，可使用向下投射照明，如台灯、立灯以及悬吊较低的吊灯，使之距离地面略近，都能收敛空间视野，饱和空间张力。

四、制造角落的微小光域

不论是沙发的角落，书房的角落，通道的角落，适当的灯具照明，能带给人舒适放松的情境。除了立灯、台灯等外加式灯具的角落利用外，在适当的地方活用嵌灯及壁灯，也是制造角落微小光域的方法。

食天下酒吧
THE WORLD CUISINE, KNOKKE-ZURI

设计师：格雷德　　　　　　　　　　　　　　　　Designer: Gerd Couckhuyt

早有夙愿设计一个全白的酒吧空间，该空间不受任何时空及地点的限制。接手本案之前，此空间由别人掌管。但是，一切皆在机缘。某天的邂逅，遇到了本案的最初业主。相遇时，偶然提及设计一个全白的酒吧，业主没有做出过多的反映。一个星期后，设计师接到了他的电话。他谈到正好有一个合作人意愿投资的项目正与设计的理念相互吻合。

几经周折，几次交谈，一切变成了现实。比利时最有名的沿海城市克诺克正好有两个空间，分别是迪吧、酒吧。

设计后的空间依然是两个空间，一主白天，一主黑夜。两者之间以一屏风相连。该屏风走势与海岸线平行。除了基本的设计理念，白色的空间内，灯光自然而然成了不可或缺的因素。模块化的灯光设计成了自然的选择。

另外的重点便是家具。沙发是定制的，椅子也是定制的。所有的家具铺陈全出自本案设计师开办的家具公司。延续的手法保证了整体上的统一。

所有音箱的扩音器全部置于墙体与天化。空间的外围增加了一道斜墙，安以推拉门，便成了卫生间的所在。

玄关如同黑色的隧道。隧道的地面和天花全为镜材，反射着两旁的灯光。整体的效果给人一种幽深但却高阔的感觉。

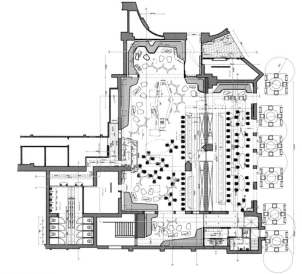

室内灯光　KTV　会所
INTERIOR LIGHTING　KTV　CLUB　063

064 流光溢彩 灯光设计
THE COLORS LIGHTING DESIGN

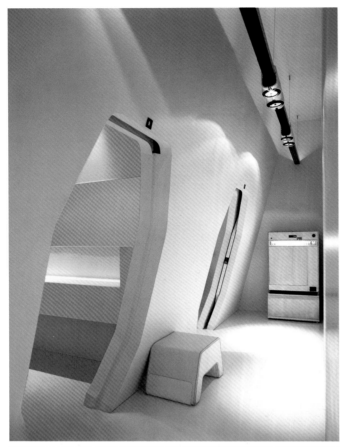

室内灯光　KTV　会所
INTERIOR LIGHTING KTV CLUB | 065

066 流光溢彩 灯光设计
THE COLORS LIGHTING DESIGN

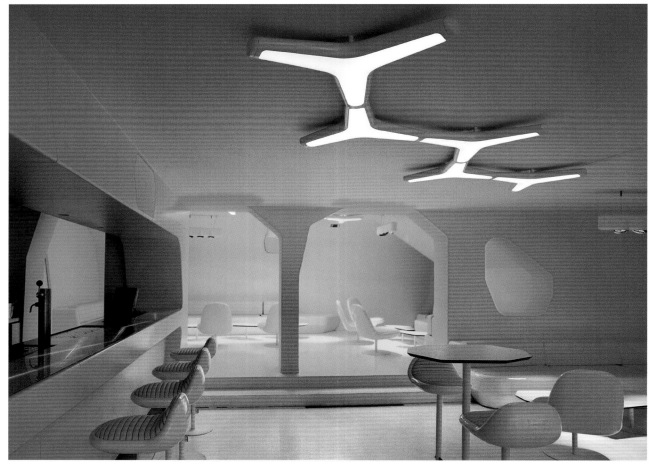

室内灯光　KTV　会所
INTERIOR LIGHTING KTV CLUB

068 流光溢彩 灯光设计
THE COLORS LIGHTING DESIGN

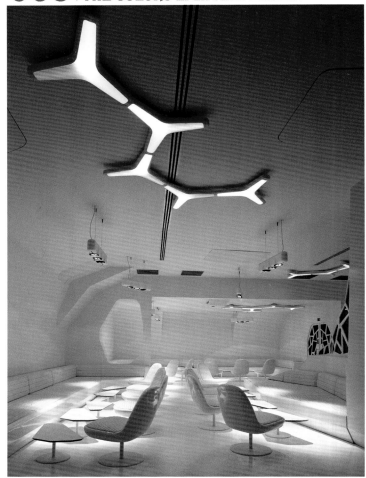

室内灯光　KTV　会所
INTERIOR LIGHTING　KTV　CLUB

About two years ago I had an idea in my head about creating a total white bar. There wasn't really any location or space determined, nor a time to execute it. Before I had created the lounge bar B-in in Brugge. In the meanwhile that bar was taken over by other people but I ran into the original owner for whom I designed it. By accident I mentioned to him the idea I had about creating the "white bar"…Initially he thought it was a nice concept but he didn't really had any intention of starting a new club, but a week later he rang me and said he met a possible partner/investor that was really interested in the concept.

There were some first meetings about my vision towards the concept and of course the location would be one of the most important aspects to create this project and after a while they started an intensive search for the ideal setting. Finally the ultimate dream location became reality. The ancient Number One disco and Bis bar that were located in the right wing of the casino building in Knokke, the most prestigious coast city of Belgium.

I drew out the major lines of construction. The main idea was to create two spaces: A day space and a night space with back to back bars divided by

a screen parallel to the coast line. Besides the basic concept there were a lot of aspects to be dealt with. Light would be a very important factor in the design because the white setting would reflect the light that was used. I created a design for Modular Lighting Instruments, the Izar.

The furniture was another great part of the project. Moon, my own company, under which I developed this concept also worked out the new created furniture. Next to custom made sofa' s the new "Zuri lounger" and "Zuri barchair" were made. The existing "Island" and "Rock Mutant" of the Moon collection were also integrated in this project.

All the loudspeakers of the sound system were integrated into walls and ceilings so no disturbing elements were seen. The bathrooms were an extension of the spacy environment using a forward leaning wall with sliding doors.

In contrast to the white club the entrance was created like a black tunnel using mirrors on the floor and ceiling that projected the light streets that were placed on the sides. The effect is an entrance with an endless depth and height.

流光溢彩 灯光设计
THE COLORS LIGHTING DESIGN

鸡尾酒酒吧
THE SMOKEHOUSE ROOM—COCKTAIL BAR

设计公司：巴斯拉一德设计工作室　　　　Design Company: Busride Design Studio

本酒吧位于印度新德里，附属于一个称为 SHRoom 的夜总会。

空间地理条件得天独厚。基地占地 1 115 平方米。袅袅婷婷，给人以有机雕刻的印象。周围风景优美，俯瞰印度宏伟景观——库特博遗产区。内里为框架式设计，以后现代主义的棱镜进行透视。

平面布局三分天下，借助设计妙手，流性、有机、梦幻般的风景融各处功能空间为一体。

太阳下的鸡尾酒酒吧显得那么纯朴、可人。晚上，借助于强烈的灯光，迷离的环境瞬间成为一个纵情的所在。周边的自然条件，内里的陈设，诸多的功能，交织于梦幻之中。模块化的酒吧高脚凳、有机形式的沙发席位十分舒适，折中的设计随时满足酒吧需要。抑或是饭店的一部分，抑或是附属的会所，但无不是各种能量的展现。

夜晚醉人的气氛，衬托着鸡尾酒酒吧的迷恋。空间任由华灯照耀，引发能量的迸发。鸡尾酒酒吧的婉转曲线、蘑菇房的灵感皆来源于墙面的质感。光照却因此变得更加有趣纷呈。

Busride Design Studio have designed a restaurant called The Smokehouse Room as well as an attached nightclub called SHRoom in New Delhi, India.

One of the most bizarre briefs we've worked with, Smokehouse Room flows organically out over 1,115 square meters, overlooking the grandest view in India, the Qutub Heritage precinct. The Smokehouse Room frames 13th Century history in postmodern lenses.

The Smokehouse Room has 3 distinct, yet seamlessly connected offerings. We've tried to create a fluid, organically growing, psychedelic landscape that melts into various parts.

The Cocktail bar introduces schizophrenia into the offering, being a pristine white space in the day, and becoming an intensely colored, constantly changing hallucinatory environment by night, which takes the trademark Smokehouse Room eccentricity into cocktails. We

室内灯光　KTV　会所
INTERIOR LIGHTING　KTV　CLUB　073

worked on a molten design form that creeps up the edges of the spaces, sculpting out functionality. The idea was to work on flexible, eccentric seating ideas, including the mound-like barstools, and the floor hugging organic sofa form, such that it creates multiple seating options, for different Bar requirements. The Cocktail bar, in layout forms part of both the Restaurant and the attached Club, hence responds to both extremities in energy levels.

The Night-time vibe, the Cocktail Bar turns into a psychedelic playground for shifting lights, to pump up the energy levels. The soft curves of the Cocktail Bar and a mushroom inspired wall texture to allow us to light interestingly.

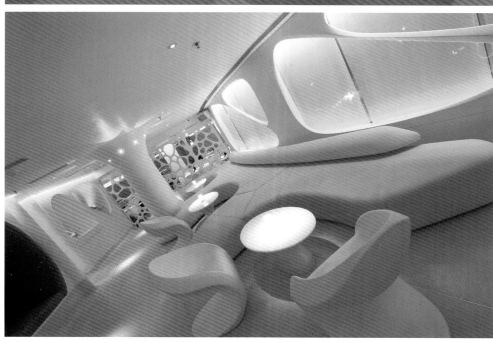

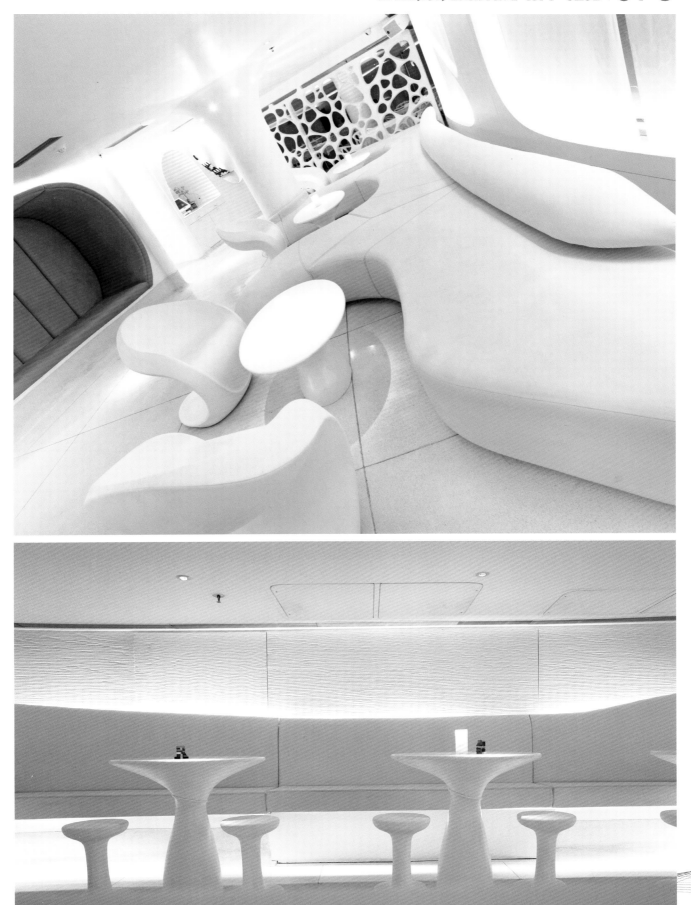

076 流光溢彩 灯光设计
THE COLORS LIGHTING DESIGN

室内灯光　KTV　会所
INTERIOR LIGHTING　KTV CLUB　077

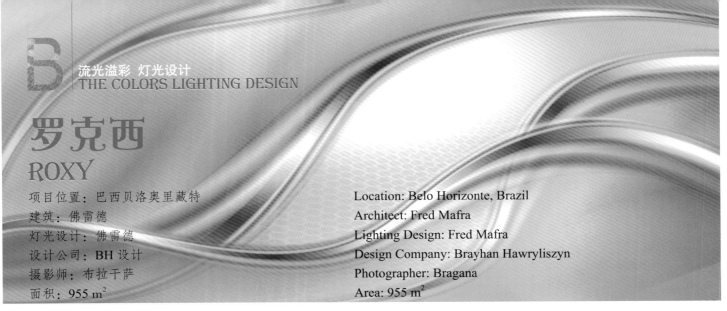

流光溢彩 灯光设计
THE COLORS LIGHTING DESIGN

罗克西
ROXY

项目位置：巴西贝洛奥里藏特
建筑：佛雷德
灯光设计：佛雷德
设计公司：BH设计
摄影师：布拉干萨
面积：955 m²

Location: Belo Horizonte, Brazil
Architect: Fred Mafra
Lighting Design: Fred Mafra
Design Company: Brayhan Hawryliszyn
Photographer: Bragana
Area: 955 m²

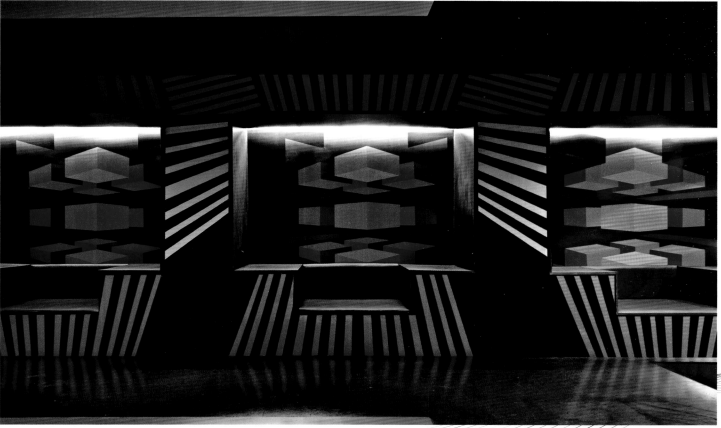

室内灯光　KTV　会所
INTERIOR LIGHTING KTV CLUB 081

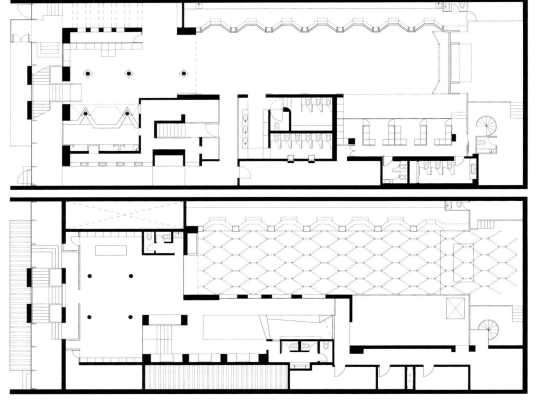

"罗克西"所在为当地夜晚最热闹的地方。955平方米的空间共有两个舞池，三个酒吧，四个VIP贵宾区。贵宾区面积可以自由界定。合而为一，便成一个超大的贵宾区域。另外还有两个休息区及带有可移动屋顶的抽烟区。

立面漆黑发亮。从里面视线可见玻璃墙。玻璃的墙面上覆盖有银黑色的釉面材质。内里结构为不锈钢制作，三角形状。玄关走道之上的每一个玻璃模块都有LCD液晶展示。总数加起来共有20块LCD。

舞台的中央，舞池尽头是DJ操作台。全方位的视线，可以看到整个会所空间。舞池30米长、6米宽，如同长长的走道。主酒吧依其势，位于一边。主酒吧设有七个嵌入式卡座。就座的人们丝毫不用担心常规直线性酒吧里常有的干扰。卡座呈梯形状，门廊模样。干式墙、声学系统的设施减少了外界声音的纷扰。而酒保却可以随时应召入内。支撑用的柱架隐于门道之内。四吨重的结构共有100多个六边形构成。每一个六边形对LED照明都是一种强有力的保护。LED灯连于可视像素映射板。U形的受光面有铝结构。该结构专门用来存放LED的镶嵌物、边口用聚碳酸酯封存。

室内灯光　KTV　会所
INTERIOR LIGHTING KTV CLUB　083

THE COLORS LIGHTING DESIGN

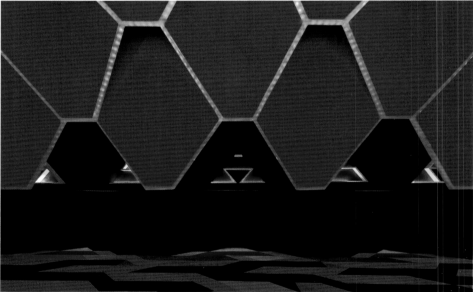

Josefine/Roxy is located in Savassi the busiest region of Belo Horizonte when it comes to night life. The place is 955 square meters of built area and has two dance floors, three bars and 4 VIP areas with retractable area(which means it can be turned into one large VIP area). It also has two lounges and smoking areas with retractable roof.

The facade is now dark black. The glass walls visible from inside were coated with dark-silver glassy material. The internal structure that holds the facade together from inside is built up in triangle shapes made out of stainless steel. Each glass module above the entrance aisle received a LCD panel. In total there are 20 LCD panels facing the outside of the club. The DJ booth which was placed at the center of a stage, is positioned at the end of the dance floor, with a full view of the club. The dance floor is a long aisle that stretches itself for 30 meters long and 6 meters wide. In one of its sides is the main bar, extended along its length. It is composed of seven bays that divide custom care and avoid crowds, which are a common problem for linear bars. The bays were framed into trapezoidal porticoes, built with double drywall and acoustic system that reduces the entry of sound, allowing the bartender listens perfectly customer's request at each niche. These gateways also hide the modular structure of the pillars supporting the struture. This four tonnes structure is composed of over 100 hexagons. Each of these hollow hexagons serve as a protective for the LED light connected to a video pixel mapping panel control. The aluminum structure in "U" receives tapes LED's inlays and then sealed with polycarbonate.

室内灯光　KTV　会所
INTERIOR LIGHTING KTV CLUB

KATAKOMB LOUNGE BAR

流光溢彩 灯光设计
THE COLORS LIGHTING DESIGN

洞穴休息酒吧

设计公司：贝特温空间设计	Design Company: Betwin Space Design
设计师：金荣光等	Designer: Kim Jung-gon, Oh Hwanwoo, Lee Gae Yeol, Kim Hyun Gu
建筑：崔银常	Architect: Choi in Chang
图文设计：YNL 设计	Graphic Design: YNL Design
摄影：李皮浩	Photographer: Lee Pyo-joon

弧形的顶、彩色的玻璃窗共同赋予了本案"哥特式的建筑主题"。几何形状的彩色玻璃天花描绘着自玄关就开始的壮丽景观。哥特式建筑中常用的拱顶、彩窗等彰显着地下的城。

金色的头骨骷髅吊灯自天花下垂，鬼魅般地洒下一地明亮。粗糙的砖墙表达的"洞穴"的影像，和谐地融于空间的神秘概念中。简约的黑白家具于传统、现代之间书写着平衡。

有趣的元素形塑了一个动感的空间。哥特式的风格表达得淋漓尽致。与众不同的特点正等着您的光临、探索。

Our space was re-created with powerful gothic architectural themes using arch and stained glass windows. Geometric patterned stained glass ceiling depicts splendid view from the entrance and such elements like arch and dome structure interior highlights the Underground City.

Gold-tinted skull pendants hang down from the ceiling and they add brightness to the space. Rough brick walls conveyed the shelter image of the Catacomb and they are well blended with mysterious concept of the space. Simple black and white furniture keep balance between modern and tradition.

Our project, Catacomb creates its dynamic space through interesting elements. Gothic styled interior was expressed powerfully, and its unique characteristics let visitors to explore the space.

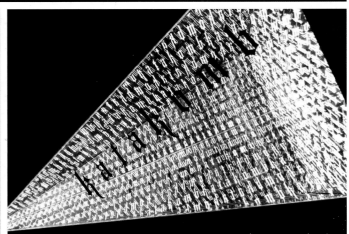

室内灯光　KTV　会所
INTERIOR LIGHTING　KTV　CLUB

088 流光溢彩 灯光设计
THE COLORS LIGHTING DESIGN

室内灯光　KTV　会所
INTERIOR LIGHTING KTV CLUB

夜飞
NIGHT FLIGHT CLUB

流光溢彩 灯光设计
THE COLORS LIGHTING DESIGN

设计公司：MODE 工作室
设计师：托斯拉夫
摄影师：3 精

Design Company: Studio MODE
Designer: Svetoslav Todorov
Photographer: 3 in Spirit

"夜飞"俱乐部是一个专门进行现场表演音乐的地方，占地1 000平方米。设计灵感源于其名称，旨在创造"漫步于或者飞翔于星空下"的景像。舞台位于中央，可以360°地与观众互动。除了三层空间有座位，阳台也有座位。VIP区凌驾于主厅之上，给人一种闺房般的感觉。

空间集功能、美学于一体。隔音的设计，加长的墙体、吸光的设施创造了夜晚的宁静。闪闪发光的球体表面，强化了中央的视角感。闪闪的发光网，如同天上的繁星。镜面的反射放大着、调整着空间的比例。

通过对各比例的巧妙修改，整个空间难以置信地有了一种童话般的气氛。精于细部，丰富于细部，不同层次的体验得到深化的同时，空间也因此具有了一种"标签"感，彰显着身份的与众不同。

NIGHT FLIGHT is a live music club that spreads on 1000 square meters. Our design was inspired by its name. We re-created the notion of a midnight walk/flight under a starry sky. We situated the stage in the center of the space in order to achieve 360° contact with the audience. The seating is arranged in three tiers and one balcony. We created a boudoir for the VIP tier that overlooks the main hall.

We had to unite function with esthetics. We set up the night scenery by using the binding effect of sound insolation—adding extra depth to the walls and absorbing all light reflections. The glittering of stars was represented by a luminous web—a spherical surface that enhances the central perspective. Mirror reflections were used by reason of multiplication and modification of the space proportions.

Our main ambition (in which we succeeded) was to use suggestion, to manipulate the propositions in order to evoke a precisely formulated effect-fairy-tale/fabulous ambiance; complexity of perception/ comprehension. Emphasizing on detail, enriching it, following the concept to a greater depth resulted in multilayer experience. Other than that, an essential part of our work was to create a unique identity, a distinguished "label".

INTERIOR LIGHTING KTV CLUB

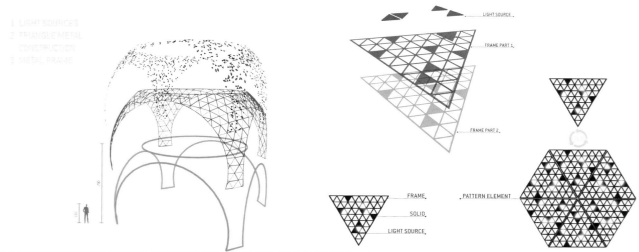

096 | 流光溢彩 灯光设计
THE COLORS LIGHTING DESIGN

室内灯光　KTV　会所
INTERIOR LIGHTING　KTV CLUB | 099

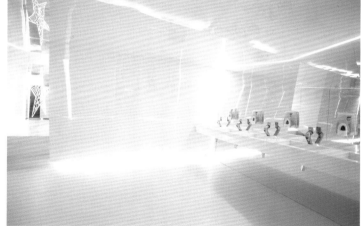

高低会所
SMOKING CLUB HI/LO

设计公司：建筑奇迹工作室
面积：200 m²

Design Company: Workshop of Architecture Wonders
Area: 200 m²

"高低会所"实是本案"天地两极"设计理念的具化。同时，本案也是对该行业不同方面的展示。现代的高端餐饮空间，可以满足食客个性化的意愿，但又兼顾客人安全、舒适的要求。所用元素，参照阿拉伯式风格、气氛。

内里设计理念参照天地之两极。借助于云景、云气、灯光, 尽现天之意向。"天堂墙"造型的完成，得益于阿姆斯特丹的代购机构。如此艺品设计，当然还包括通过唯一后窗而入的太阳光线。

自玄关开始的有机玻璃，引导着客人的行进。立体的玻璃云形与钢构，实是销售操作台与屏风。销售台、屏风之后即是自动咖啡售货机。玻璃、钢构、皮革用材的巧妙使用，不仅美观，还有效地消减着外在环境所带来的不利因素。

"地"之意向表达则更为直接。荷兰东部有一些黑乎乎的房间，人可以在此停留数天。"地"即意向之展现，正是借助于如此理念。只不过，采用了"阿拉伯"式的主题。皮质的沙发，簇拥在周围，以"摩洛哥"的地产高端手工饰品作为点缀，形成"天"之意向，与背光的铝板墙、黑色的玻璃盒形成鲜明的对比。铝板墙下、玻璃墙中，镶嵌着自动售货机、电视机。

设计师对厕所和储藏进行隐形设计。窄窄的通道、明亮的水槽、便触式小便器，由本土设计大师自行设计，升华着此处与众不同的体验。

设计的主要目标原本就是为了记忆的创造。不同阶段的享受全部是为了食众。这是设计的工作，也是本案设计的乐趣。

The name Hi/Lo visualizes our bipolar concept of Heaven & Hell that lies at the root of this interior design. Our brief at the time was to design a contemporary high-end hospitality space in which one smokes weed and hash in a context of security and comfort using elements that make reference to an Arabic atmosphere.

The interior concept was derived from the original name "destination", considering there are only two ultimate destinations we decided to make them both, heaven and hell. For "Heaven" we were inspired by the idea of floating in a cloudscape, airy, light and high. The one off heavenly Wall Art created via the Shop Around agency in Amsterdam supported this fully, including the sunrays radiating from the only back window that we transformed into the sun.

We designed large organic glass elements for the entrance to control the client access and added cubistic clouds of glass and steel to create a sales counter and a screen to hide the vending and coffee machines. By using glass, steel and leather as main materials we created a resistance to the many hot butts and roaches.

"Hell" is as expected much more direct, it was inspired by the opium dens of the east, those dark rooms where one stays for days under the influence of that mystic substance. Here we introduced an Arabic theme with leather couches to lay on surrounded just like in "Heaven" by high-end design products mixed in with local Moroccan handy crafts and contrasted by the back lit exploded aluminium wall panels and black glass casings for vending machines and TVs.

We hid the toilets and storage in a black box finished with a Tadelakt coating. The toilets enhance the experience level of someone who is stoned by introducing narrow passages, illuminated sinks and the kisses urinal designed by local designer Meike van Schijndel.

We have once again been able to create a space that supports our main goal of creating memories. We love being part of the experience that customers will have enjoying this environment in their various stages of getting high.

室内灯光　KTV　会所
INTERIOR LIGHTING KTV CLUB 103

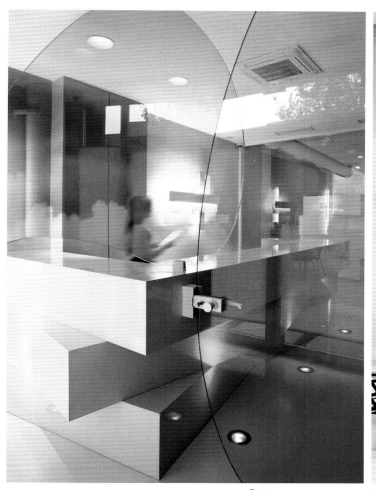

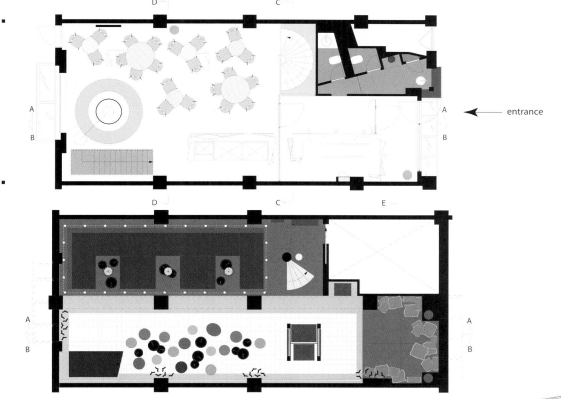

entrance

INTERIOR LIGHTING KTV CLUB | 105

流光溢彩 灯光设计
THE COLORS LIGHTING DESIGN

莆堡酒吧
POPO CLUB

项目位置：广东惠州	Location: Huizhou, Guangdong
设计公司：深圳市新冶组设计顾问有限公司	Design Company: Shenzhen Newera Design Consulting Co.,Ltd.
主案设计：陈武	Chief Designer: Chen Wu
参与设计：吴家煌、代浩、张春华	Participate: Wu Jiahuang, Dai Hao, Zhang Chunhua
面积：3 000 m²	Area: 3,000 m²

一座建筑物的个性，从外立面上就可以看出。建筑物的价值观和生活品位，也通过外立面豁然张显。眼前这个建筑物的绚烂和活力，让人想起电影《了不起的盖茨比》，奢华的场景令人大饱眼福。

一个璀璨的世界，一场声色犬马的华美盛宴。片片金钱做瓦，架起一座寻梦的空中楼阁。上流社会奢华浮靡的躁动质感，精致的光效，耀眼的水晶灯饰，纸醉金迷的奢华，雕刻出令人叹为观止的质感。

当看到POPO CLUB这栋色彩斑斓的建筑物，不知不觉便生出由衷的感动，犹如古堡华府场景重现，但在极致古典奢华的外壳下所包容的却是一条盘踞前沿娱乐文化的巨龙。应该是对生命有着非凡热情的设计师，才会首先在心里造出这般五彩斑斓、热情洋溢的空间，从而给更多的普通人的生活带来多样化的动力。

2013年12月完工的POPO CLUB，实用面积3 000平方米，楼上楼下共两层，一楼为酒吧大厅空间，二楼分布豪华包房。娱乐也需要筋骨与质感。外立面的工业风格用料，包豪斯式的现代主义风格硬装，桥接奢华精致的琉璃堡垒，构成绝佳的视觉盛宴。室内空间形式以豪华的火车车厢呈现，个性十足的重工硬装空间点缀飘逸的炫彩元素，为新时代的娱乐贵族量身定制高质感的欢愉场所。

POPO CLUB是由深圳新冶组设计顾问有限公司所设计的。一如许许多多优秀的设计团队，新冶组也是一个致力于追求卓越设计的团队，希望通过强有力的概念到清晰的建筑形式来实现他们的建筑物。这个设计团队的作品，建筑形态上以不定型几何体为主要形式。但在组合形式上颇有些创新的方式，既有西方工业时代的稳重感，又不乏新冶组特有的求新求变的气质。他们的作品活力通常由色调来控制。颜色在空间中的对比、不拘一格的材料使用手法，使作品也呈现着稳重而活泼的气息。

作为惠州最前沿的新潮酒吧，POPO CLUB几乎是惠州酒吧形象的绝对表征。于空旷废墟地段平地而起，划破惠州入夜后的沉静，带来喧嚣与激情。POPO CLUB给了人们一次短暂的旅行机会，在这个承载着欲望的列车上，好奇的眼睛为同类的身体而张开，并给予匆忙的一瞥。你的眼睛彻底被眼前的美丽、奢华所折服，在完美感、奢华感与空间感中享受着轮番轰炸般的震撼。

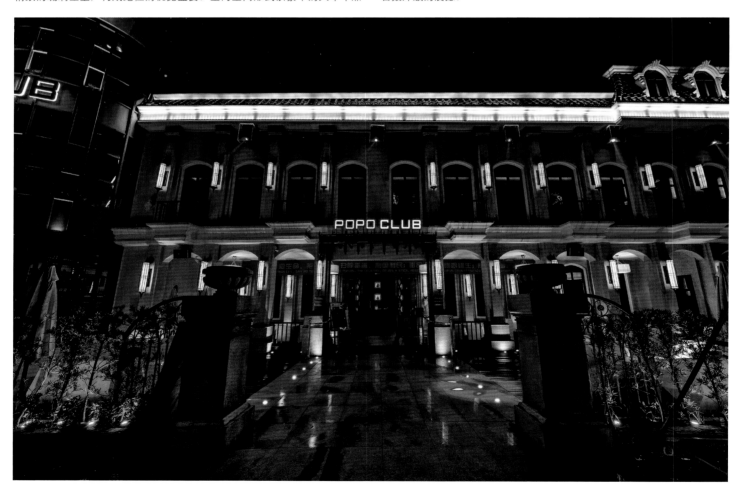

From the external, we can know the personality of a building, while from its facade, we can get perception of its value and life value. With its brilliance and vigor and vitality, the building is reminiscent of The Great Catesby, whose luxurious scenes are bound to glut people's eyes.

A bright and splendid world this project makes where to offer gorgeous feast. Tile seems to have been made of pieces of note to build up a castle in the air via which to seek dream. Its quality exclusive to high rank, its delicate lighting effect, and its shiny luxury, altogether shape texture that you feel nothing but marvelous for it.

The sight of this colorful building generates a kind of immediate touch. Scenes similar to what usually takes places in castle, but inside the fashion coat is a huge entertainment cultural dragon that lies on the leading edge.

Finished on December, 2013, the volume of POPO Club has over 3,000 square meters over two floors. The ground floor is used for bar lobby, while the 1st floor for private boxes. The materials involved for the skin is of a strong industrial style, the decoration is very modern and the bridging of glass is quite luxurious and delicate, all making a superexcellent visual feast. The interior is in form of railroad car with element of heavy industry to embellish the color to specially-made for the guests.

Designers working with a good team which is aimed to seek beyond what is has been and achieve their mission by presenting buildings with concept power and clear building. Works by the company is in the pursuit of novel creation instead of regularly geographical form but with stead sense of the west industry and temperament of novelty aspiration. The vigor of their project is often done by hues, where color contrast and materials not sticking to one pattern also bring out a sense both calm and vivid.

As the most pioneering club in Huizhou, POPO Club almost makes an absolute representative that rises above ruins and flashes in night to offer enthusiasm and passion. Here provides a chance to get a short journey, makes a train as a carrier of desires, and presents eyes that submit to the beauty and luxury in being shocked by perfectness and senses extravagant and spatial.

马来西亚 SOJU 吉隆坡店

SOJU

项目位置：马来西亚吉隆坡
设计公司：深圳市新冶组设计顾问有限公司
主案设计：陈武
参与设计：张春华、吴家煌、代浩
主要材料：土耳其灰大理石、柯伊诺尔（水珠面）、黑白根大理石、热带雨林棕大理石、松香玉透光石、山东白麻、美国灰麻、紫竹文化石、艺术漆、氟碳漆（肌理面）等
面积：1 100 m²

Location: Kuala Lumpur, Malaysia
Design Company: Shenzhen Newera Design Consulting Co., Ltd.
Chief Designer: Chen Wu
Participate: Zhang Chunhua, Wu Jiahuang, Dai Hao
Main Materials: Marble, Cultured Stone, Paint
Area: 1,100 m²

空间布局始终是与酒吧经营业态相挂钩的，与其他酒吧不一样的是，SOJU 既有浪漫闲适咖啡吧，也有动感刺激的酒吧，以适合不同人群和不同需求人群的消费。

咖啡吧位于 SOJU 四扇大门后的区域，照片墙与撞色是这个空间的主题，黄色的墙面与紫色的餐桌椅相映成趣，相框中的大幅音乐人照片与照片墙上贴着的各种记忆瞬间相互对望，木质的吧台与地板与由文化石构成的墙面相互映衬，眼神不由自主地注意到了墙面上随处可见的英文LOGO："SOJU"、"PARTY"、"CLUB"、"SHOW"……

在通过由绿色射光交织的通道后，柔和明亮的光线变得暗淡暧昧起来，深色色系成为了空间的主色调，在幽长的走道两侧，具有古典风格的木质墙面中却夹杂着类似钢筋铁条一样的栅栏，猎奇的心理油然而生。

步入门内，别有洞天。舞台，整个空间的中轴线就是长长的 T 台，对面则是一面大大的 LED 屏幕，"WELCOME TO SOJU!" 两侧密集的射灯从不同的角度，照亮了这本应黑暗一片的白色路径，仿如星光大道，当 DJ 台上的音乐响起，光、影、声、乐，舞动的舞者与四周散台上舞动的人群们一起摇摆，要把这属于尘世的烦恼统统抛在脑后！

靠近场地边缘的沙发区域则显得光线暧昧，宽大的皮质沙发与金属质地的栏杆围成了一个个卡座区域，不时有射灯飘来，颜色素雅却色彩斑斓的沙发区域与闪着金属色彩的茶几时隐时现。

多元化的空间元素，需要用多元化的材料来凸显，设计师的选材几乎遍布了整个世界，从土耳其灰大理石，到热带雨林棕大理石，从山东白麻到美国灰麻，从艺术漆到氟碳漆……

冲突与对话始终是整个空间的主题，设计师的创新在于将不同元素的装饰材料有机地进行了融合。

夜店的消费呈现出的多元化，决定了不同酒吧的定位以及经营业态与服务模式，设计师在设计 SOJU 酒吧的时候，更多考虑的是用不同的设计风格来满足酒吧业主对于消费模式的把控，在投入运转之后的 SOJU，满足了吉隆坡市中心消费族群对于夜生活的要求，自然大受追捧。

Despite its usual layout, SOJU integrates cafe bar and bar into one to meet needs of different customers and people.

Behind the four doors of SOJU is the cafe bar, where picture wall and contrasting colors are the theme. The yellow wall and the purple dining table and chairs are set off with each other. The large pictures of musicians are corresponding to those as carriers of different times. The wood of the bar counter and the flooring is set against the cultured stone of the wall. Everywhere are English words on the wall, like LOGO" SOJU", "PARTY", "CLUB" , "SHOW", etc.

The corridor is cast green by spot light; then light becomes softer and darker, and dark colors are beginning to dominate the space. Flanked with the long aisle are classical wooden walls, between which is a fence like bar iron, generating your curiosity to seek novelty.

The inside really makes quite a different world. A long T stage serves as the central axis, opposite to which is a large LED screen, with WELCOME TO SOJU on it. Light from all directions by numerous spot lights illustrates the path, making an avenue of stars. When

DJ comes, light, shadow, music and dancers are swing and swaying, keeping away all the hustling and bustling.

The soft area at the space edge looks ambiguous, where large leather sofa and metal railing embrace a deck. When the light comes by, sofa of colors and metallic tea table come into sight, and they goes invisible as the light goes away.

Diverse spatial elements needs different materials to set off. Those used in this space are from all over the world, gray marble from Turkey, brown from the Rain Forest area and all you can name.

Conflict and dialogue are always carried out throughout as the theme. With novelty and creation, elements of different kinds are combined organic.

The pluralistic consumption in night entertainment avenues determines its changeable positions, business and service. Design for this space lays more on styles to meet the owner's demand of consuming model. Due to its catering for the preference of the customers, it's no wonder won great affection from the market.

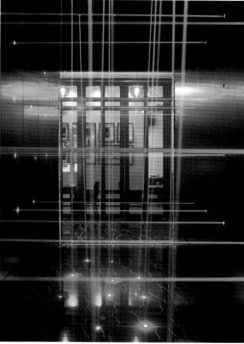

流光溢彩 灯光设计
THE COLORS LIGHTING DESIGN

南京 Enzo 酒吧
ENZO BAR, NANJING

设计公司：灵感集团
面积：1 073 m²

Design Company: Inspiration Group
Area: 1,073 m²

有音乐，有酒，还有很多的人，这是传统观念中人们对酒吧的印象，而谁都无法阻挡自己的内心对酒吧的向往与依赖。每当夜幕降临，都市中灯影闪烁，Enzo 便是金陵城中最闪亮的夜名片。

围绕"玩乐"、"互动"、"愉悦"三大设计需求，将功能与装饰、科技相结合，令 Enzo 独具特性。颠覆常规酒吧的动线分区，在前厅处增加与玩乐相关的商品售卖区域，各类时尚产品方便客人进行主题派对，同时还配有 Lomo 相机出租服务，让每个开心的瞬间能够即刻呈现，并永久珍藏。让 DJ 与客人、舞者与客人以及客人与客人之间开心互动，除了以半包围布局加散台的形式强调感观交流，还将场内卡座全体解放，扩大活动空间，让客人可以根据人数、玩法自由组合座位形式，制造活跃气氛。当音乐转场舞台会跟着转变，动感的 LED 屏幕前移动门打开，舞台延伸至屏幕，舞者惊艳出场，通过地面 LED 的动感处理，辅以淋水式舞台设计，平面、立体双重结合，全场每个角落都可享受丰富且有激情的视觉体验。

整间酒吧以暖灰色主调铺陈简约时尚的设计手法，营造出专属都市时尚的 Lounge 格调，加入紫色、金黄、湖蓝、火红的光影，让空间层次更为立体与丰富。从视觉、嗅觉、感觉、听觉、触觉五感营建尊贵的品位与魅惑的气质，令 Enzo 成为中外潮流人士的聚集地，同时也成为南京城中娱乐新地标。

室内灯光　KTV　会所
INTERIOR LIGHTING KTV CLUB

室内灯光　KTV　会所
INTERIOR LIGHTING KTV CLUB

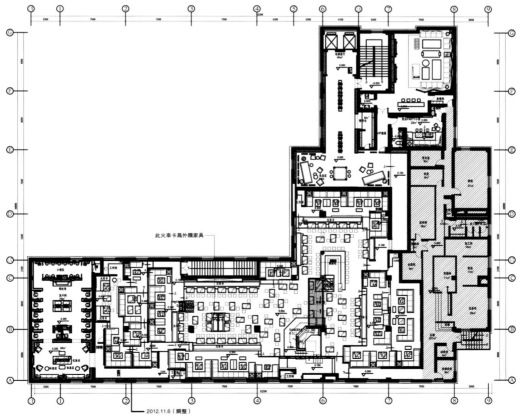

In a bar, there is always music, alcohol and people. Such is the usual impression people have for bar while nobody can resist their inner desire for bar. As night set in, Enzo makes a name card that glitters most in the metropolis of Nanjing.

Centered on "fun", "interaction" and "pleasure", here combines decoration and science and technology with its unique character. The stereotyped spatial division in a bar is reverted with part of the vestibule used for item sales. All fashion items bring convenience and ease for guests to have theme party. The Lomo camera ready for guest to rent makes record of memorable moments. DJ, dancer and guests can interact as freely as they want. The semi-closed space and seats enhance the communication between while the dock is completely opened to maximize the activity space, sustainable for seat arrangements. When music comes on, LED screen opens, to which dance flooring to hold dancers then. LED is treated dynamic. The water flowing dancing flooring either of plan or 3D, allows for visual passion and enthusiasm everywhere.

Warm-gray in the tone, the space creates a lounge style exclusive to the urban fashion. The appearance of purple, gold, lake blue and fire red enriches the space. Five senses are all employed for shaping taste, and consequently the bar of Enzo makes a place to get together of people home and abroad, erecting a recreational landmark in Nanjing.

B-ONE 酒吧会所
B-ONE LOUNGE CLUB

流光溢彩 灯光设计
THE COLORS LIGHTING DESIGN

设计公司：齐合作设计
设计师：金智浩
参与设计：李东元、金炯锦、做妍、郑惠利、朴赞联
摄影师：伊蒂弗斯
面积：958 m²

Design: Chiho & Partners
Designer: Kim chi ho
Participate: Lee Dong-won, Kim Gin-hyoung, An Do-yeon, Jeong Hye-lee, Park Chan-un, Oh So-jung
Photographer: Indiphos
Area: 958 m²

位于梨泰院中心位置，B-ONE 酒吧会所无疑是自由与青春的象征。空间以变化、发展的内涵，重塑地下室、团队文化、音乐等象征性元素，表达着一体独特、体验般的感性，实现着汉密尔顿酒店地下空间的拓展。通过各建筑元素的重叠，如拱门、立柱、墙壁、天花自由、粗糙的质感等，整个空间弥漫着一种稳重。是出于必要，也是出于实用，三个原本独立的区域，如同盒体般实现着彼此之间的连接，同时以各自不同的理念，激发着对空间的无限遐想。

通过长长的阶梯，猛然给人一种梦幻般的感觉。整齐、有序的立列排列，红绿色调的对比致人以国度、力度之感。绿色的地砖，令人耳目一新。制陶艺人手工制作的地砖顷刻间实现了 B-ONE 给人的第一印象。

VIP 区域，红色的热情有着透明的纯粹，A 区、B 区之间可以相互观看。VIP 区域，通过所产生的色度、视觉框架，红色的强化玻璃看起来动感十足。涂面材料，集粗犷、细腻于一体，用材的重复与不同视角起着不同的效果，B 区整体有了一种时髦的感觉。空间古旧的陈设，用其曾经的沧桑、时光的印迹，讲述种种"故事"的内涵。盒状的 VIP 包间，自然抬升于地面，格状修饰，LED 照明。相互交叉的两个舞台，从视觉上、功能上中和整个空间。

C 区有着光滑的金属装饰，灿烂的金色、黑色图案。各种用材颠覆着其固有的质感，强调着其本身的存在，形成富有想象的图案。精典元素、LED 显示的功用集合，是实验，更是创意。整体看，空间依然和谐。

不同概念，不同气氛，但三个分区依然和谐共生。出于实际需要，或开或合，方便相互呼应的同时，又创造着令人敏感的变化。

Located in the center of Itaewon, this space is B-ONE Lounge Club to symbolize freedom and youth. Expanding and moving to the basement of Hamilton Hotel, it had to reinterpret the symbolic elements of club such as basement, party culture and music with a meaning of change and development, and express the more experimental and unique sensibility. With this reason, designer planned the gloomy atmosphere of basement to overwhelm the whole space by overlapping the architectural elements such as arch, pillar, and rough texture of wall and ceiling freely. Client wanted to use this space partially as necessary in order to use it practically. Three sections are severed completely and connected as one from case to case, and stimulate the ceaseless curiosity on the long and narrow journey with the different concepts each.

Section A, where we meet with the dreamlike image of club first at the moment we arrive at the basement through the long stairway, gives the sense of depth and tension to the space through the rhythm from the pillars placed regularly and sequentially and the strong color contrast of green and red. Floor covered with tiles in green keeps the feeling of refreshment, and the tiles made by a potter by hand completes the first intense image of B-ONE Lounge Club.

Section A and section B can look in each other through VIP Zone of red and transparent mass. Scenes seen through the red tempered glass of VIP Zone look more dramatic with the optical illusion by the color and visual change of frames. Section B features the effect produced by the contrast of rough and refined finishing materials and the unexpected effect from the diverse visual angles of repetitive elements, and tells the story through the old things with rough texture and trace of times. VIP booth in the shape of box decorated with lattices creates an intense scene by raising the level of whole floor and finishing it with LED lighting. Two stages placed across the section B play the role to neutralize the whole space visually and functionally.

Finally, section C is completed with the dramatic harmony of glossy metal decoration and splendid pattern on the golden and black colors. Elements reversed by the property of

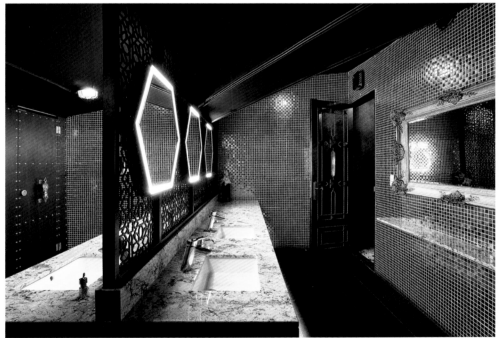

materials emphasize the sense of existence each other and form the imaginative images. In particular, functional combination of classical elements and LED screen brings the more experimental and creative change.

Like this, three sections have the different atmospheres and concepts but at the same time harmonize naturally one another. They are opened and severed flexibly according to the functional necessity of differentiated club management for the smooth communication with the clubers free and sensitive to the change.

FAT LADY DISCOTHEQUE

胖夫人迪斯科舞厅

THE COLORS LIGHTING DESIGN

设计公司：M & Y 工作室
设计师：安蒂
参与设计：基莫、埃里卡、卡伽耐
摄影师：图奥马斯
面积：1 300 m²

Design Company: M & Y Architect Studio
Designer: Antti Moisala
Participate: Antti Moisala, Kimmo Yl-Anttila, Erica Kivel, Kimmo Karjunen
Photographer: Tuomas Uusheimo
Area: 1,300 m²

"爱酒店"，当地夜生活最为繁华的地区。如今，借助于设计妙手，空间华丽转身，"胖夫人迪斯科舞厅"就这样走到了大家面前。主舞厅彰显数字化的多层焊接。吧台与吧桌尽多层次糅合。电脑数控制作的麻纸板，以3D弯曲的姿态呈现。3.6米宽，25.2米长的LED网状视频天花，以30根缆线连接，视频、光照的3D效果就这样永远呈现。DJ，音乐控制的升降台、照明、显示屏尽情展示着"胖夫人"的身份特征。

空间所呈现的双层弯曲制作工艺借助于严格参数和自动化的流程。图纸全部通过AUTOCAD制作。高科技的精工手法，确保着一个与众不同的"胖夫人"。

A total renovation was done for the previous hotspot of the city nightlife Love Hotel, now turned into Fat Lady – named not least according to the new interior design elements. They consists, among other things, of digitally fabricated multilayered wall structure in the main dance hall and several bar counters and tables incorporating layered and cnc-milled board structures resulting in 3D curvatures. As a top of the Lady there is a 3.6 m wide and 25.2 m long LED video net ceiling, which movement is controlled by 30 wire lifts, producing everchanging 3D formations with video and lighting effects - a continuously parametrically controllable spatial element. Lifts, lights and video screen are all controlled by DJ as well as music.

Most of the double-curved forms were worked in Rhino, also utilizing Grasshopper parametric control and automatization routines. Drawings were exported to Autocad for collecting and finishing the drawings before sending to production.

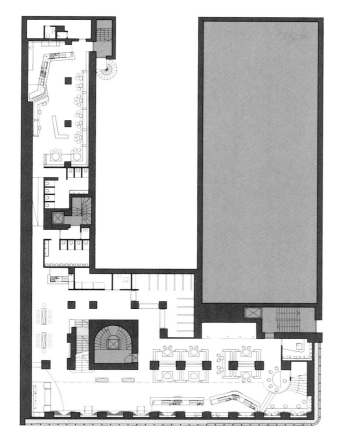
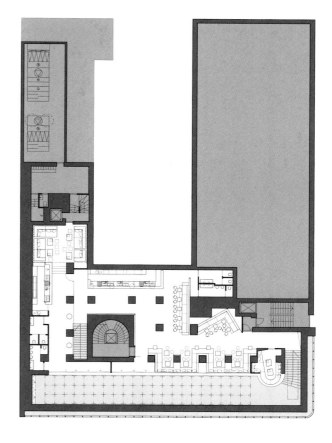

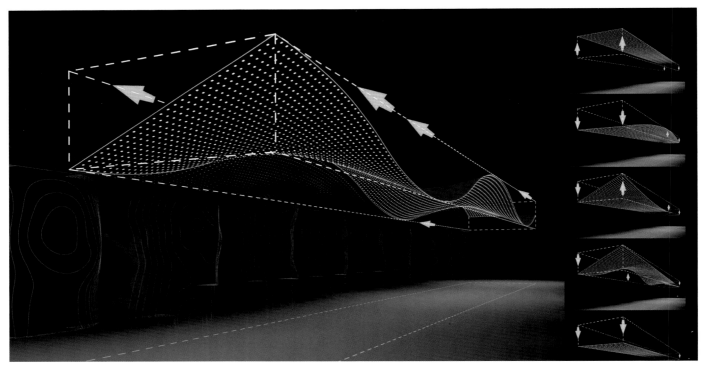
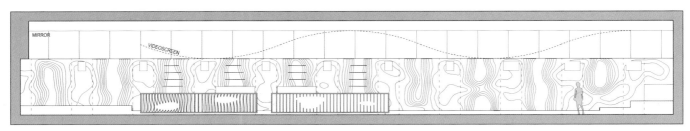

流光溢彩 灯光设计
THE COLORS LIGHTING DESIGN

重庆丽芙酒吧
LIV SHOW BAR

项目位置：中国重庆
设计公司：深圳市新冶组设计顾问有限公司
主案设计：陈武
参与设计：张春华、吴家煌、代浩
主要材料：柯伊诺尔、斑马石、鱼肚白大理石、浅米白色水磨石、艺术漆、氟碳漆、钛金
面积：925 m²

Location: Chongqing, China
Design Company: Shenzhen Newera Design Consulting Co., Ltd.
Chief Designer: Cheng Wu
Participate: Zhang Chunhua, Wu Jiahuang, Dai Hao
Main Materials: Marble, Terrazzo, Artistic Paint, Titanium
Area: 925 m²

2013年12月，高端娱乐品牌LIV SHOW空降重庆江北，绚丽的灯光与高品质的材料共同构建了一个诱人的空间，时尚的环境与性感的线条带给人们最亲密的接触，让人放松，沉溺其中。

LIV show club处于重庆市江北区CBD娱乐业态核心位置，重庆娱乐向北发展归属地，拥有超过2 000平方米娱乐现场。会场综合了宏观的环境和优越的地理位置优势，定位高端夜店俱乐部、国际、品质、健康、互动是LIV show club发展的出发点，坚持主题派对为核心竞争力。

LIV show club在美国迈阿密派对夜生活的基础上延续时尚、潮流、火爆、新奇、奢侈、高能量，以高端主题派对及全球化全空间为核心，打造差异化明显，核心竞争力强劲的高端娱乐品牌。设计师在接到设计委派后，首先帮助业主对其经营业态进行了规划建议，室内设计方向则根据业态需求确定。

设计师陈武以巨型鸟笼的概念来打造这个空间。穹顶式镂空鸟笼结构令大厅空间通透而高大，笼架上包裹LED灯条，黑暗中漫射出绚丽的光彩，形成一个氛围浓厚的电音场。在这个高度对称的空间中，稳重与突破并存。内部以新古典风格为基调，超炫丽的现代灯光与极具古典艺术气息的主题融合在一起，矛盾的冲突在这里被转化为艺术的魅力，加上灯饰与绚丽灯光的映衬，空间氛围更具魅力。

德国进口LASER WORLD激光灯结合先进的3D MAPPING技术，呈现3D立体背景舞台空间，打造颠覆性的舞美效果，实现科技与娱乐的完美融合。夜色中狂欢的人们流连、沉醉于光影音效的艺术魅力，体验前沿的科技与设计美感。

室内灯光　KTV　会所
INTERIOR LIGHTING KTV CLUB

December, 2013 saw the landing of LIV SHOW on the north bank of River in Chongqing, making a high-end charming project of lighting and exquisite material, where people can be simmered in a most intimate and relaxing ambiance with its fashion setting and lines.

A space it is located in the core area of CBD with an area of over 2,000 square meters. It comprehensive surroundings and good geography as well as its expensive position of a global night entertainment avenue present a design that's centered on international quality, health and interaction while taking theme party as its most competitive advantage.

Based on the night life of Miami, LIV SHOW club carries forward a high, and energetic trend of fashion just to create a unique recreational spot. Suggestions on business management is followed by interior design.

A huge bird cage is turned to bring out a final result where the bird dome is large and transparent. On the bird is decorated with LED lighting belts, glistering in dark to bring forward a rich electronic. In a very symmetrical space staidness and conflict coexist. The interior is dominated with a neo-classical style, in which lighting modern and atmosphere very artistic and classical are combined, conflict is shifted into artistic aroma, decorative lighting and brilliant light enhance the spatial attraction.

Out of the Laser World lighting imported from Germany as well as 3D Mapping technology is a 3D stage that reverts the usual effect by fusing perfectly science and technology and entertainment, so that people can be lost here and feel everything but unwilling to leave only for the leading appreciation of technology and design aesthetics.

B 流光溢彩 灯光设计
THE COLORS LIGHTING DESIGN

Nova 酒吧
NOVA

设计公司：何宗宪设计有限公司
设计师：何宗宪
参与设计：林忠明
摄影师：Dick
主要材料：镜面、清玻璃、砖、石材、镜面不锈钢、镜面马赛克、油漆、壁纸、皮革、窗帘
面　积：373 m²

Design Company: Joey Ho Design Limited
Designer: Joey Ho
Participate: Michael Lam
Photographer: Dick
Main Materials: Mirror, Glass, Tiles, Stone, Mirror Stainless Steel, Mirror Mosaic, Paint, Wallpaper, Leather, Curtain
Area: 373 m²

全案设计的关键灵感是店名 Nova。Nova 意喻新星，当新星爆发的时候，会突然增加亮度，绽射能量，设计师以"爆发及炫耀光芒"概念诠释全案造型，打造出科幻奇想的视觉效果，借以描绘空间、时间与能量所交融下的耀眼时刻，透过这座奇想式的宇宙时空，带领宾客从现实生活短暂抽离。故事发生在遥远的星系里，一颗新星爆发的瞬间，丰沛能量释放而出，此时受到能量撞击，星球碎片与能量线散射于宇宙之中，交织成瑰丽而奇幻的景象，也开启了令人兴奋的新篇章。宾客就像是游走于星际之间的宇宙公民，见证了这个耀眼时刻，参与着热烈的庆祝活动，享受欢乐气息。

Its name of Nova offers its key design for the whole project. Nova refers to a new star that releases a tremendous burst of energy and suddenly becomes extraordinarily bright in the sky. With a concept of "Outburst and Ray" carried out throughout, a visual effect depicted in science fiction is presented to make a vivid count when space,

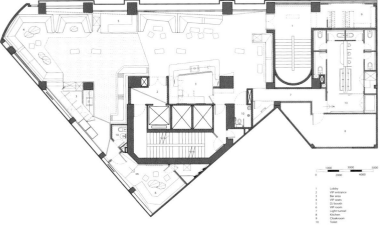

1. Lobby
2. VIP entrance
3. Bar area
4. VIP seats
5. DJ booth
6. VIP room
7. Light tunnel
8. Kitchen
9. Cloakroom
10. Toilet

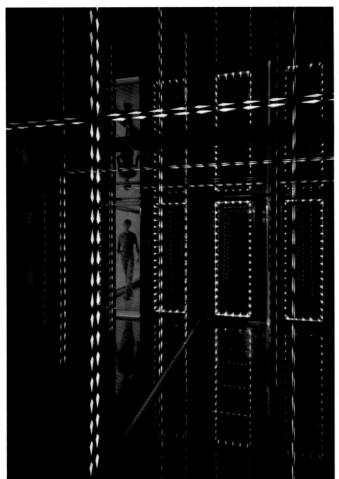

室内灯光　KTV　会所
INTERIOR LIGHTING KTV CLUB

time and energy are in communication. A cosmic space deprives the space of reality and leads onto a star remote but ready to burst out to release its energy with chips, thereby making a fantastic mirage to start a prelude of excitement, by which guests feel like a citizen in the universe when witnessing the glittering moment in participating the ceremony in a happy atmosphere.

江西永修东方玛赛音乐会所
MUSIC CLUB

流光溢彩 灯光设计
THE COLORS LIGHTING DESIGN

设计公司：福州维野餐饮娱乐空间策划设计机构
材质：黄龙玉大理石、护墙板、中纤板雕花、马赛克拼花、金茶镜
面积：2 300 m²

Design Company: Fuzhou Weiye Design
Materials: Marble, Chair Rail, MDF Carving, Mosaic Parquet, Tawny Mirror
Area: 2,300 m²

新古典是参考过去，透过历史去找灵感，创造出一种古今皆宜的新风格——新华丽古典主义。

当客户带我到原始现场看到一个闲置了两年的粗糙的中式建筑时，我发愁了，初想定位当地最高端夜总会，但将其外观全部包裹起来，这样造价会高出很多。苦思几日无果，只能借助这带有东方情怀的建筑，以浮夸的装饰去凸显这些元素，赋予这些装饰新的特质，呈现出完全不同于过往的新面貌。当外观手绘方案出来感觉还不错，于是贯穿于室内将这手法进行下去。

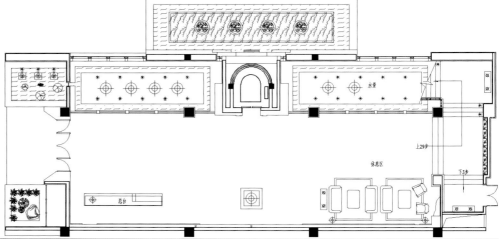

Newly-built antiques with historical inspiration can make a style popular in both ancient and modern times. That is neo-extravagant classic.

The space makes a project based on a rough Chinese building abandoned for two years and planned to be positioned as the most sophisticated club. Due to the high budget if the whole body is coated, commercial decoration is finally used to set off the oriental feelings to offer the effect new traits. A kind of face comes completely differentiated from the past. The hand-painted skin is carried out through the interior. With reference and inspiration from the old, such a style is thus accomplished.

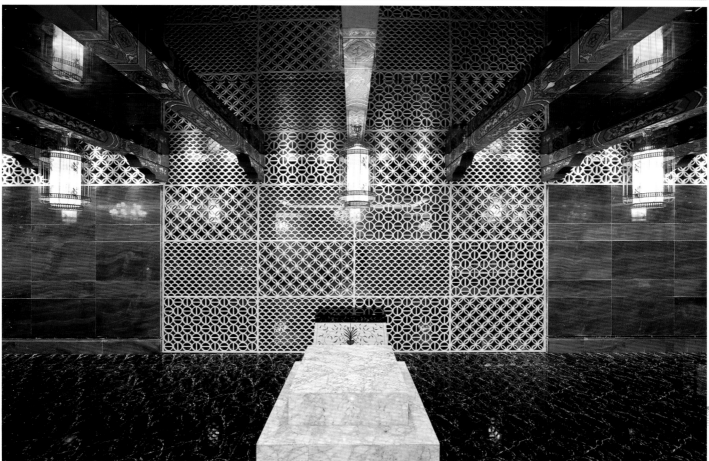

158 流光溢彩 灯光设计
THE COLORS LIGHTING DESIGN

160 流光溢彩 灯光设计
THE COLORS LIGHTING DESIGN

流光溢彩 灯光设计
THE COLORS LIGHTING DESIGN

南通皇后
CLUB QUEEN, NANTONG

设计公司：深圳市新冶组设计顾问有限公司
设计师：陈武
参与设计：代浩、吴家煌、张春华
主要材料：格棱玉大理石、奇幻彩虹大理石、啡金（水珠面）、冰花玫瑰大理石、水纹银大理石、仿古镜、玻璃水滴砖、六角砌面玻璃砖、艺术漆、双层金属帘
面积：2 390 m²

Design Company: Shenzhen Newera Design Consulting Co., Ltd.
Chief Designer: Chen Wu
Participate: Dai Hao, Wu Jiahuang, Zhang Chunhua
Main Materials: Marble, Mirror, Glass Brick, Artistic Paint, Double-Layer Metal Curtain
Area: 2,390 m²

CLUB QUEEN 品牌诞生于澳门，首家皇后酒吧 Queen's bar 诞生于 1998 年澳门珠光大厦。2011 年内地第一家 CLUB QUEEN 旗舰店在杭州惊艳绽放。2012 年潮袭郑州，火爆中原。历经两年的精密策划，2013 年底登陆江苏南通，续写传奇。在葆有纯正品牌血统基础上进一步实现转型升级，启动终极夜店模式的梦想。

在南通 CLUB QUEEN 的设计中，设计师意图打破建筑传统，改变人们对空间的看法和感受，创造拥有神奇的飘浮感的未来建筑。前卫的抽象造型，营造建筑物与地面若即若离的轻盈状态，达到一种海市蜃楼的效果。锐角尖顶、流动丝巾一样的长弧曲线，给人带来前所未有的视觉冲击力。在这里，空间就像橡胶泥一样，任由改变形状。

设计师将现实转化为丰富交缠的世界，尝试捕捉不断变化的能量。将多向度透视、快速移动而强烈的造型和科技性的架构杂糅整合，构成多义性的意象表现。大胆运用空间和几何结构，创造出超出现实思维模式的新颖作品，也让我们看到了城市生命力的喷薄和流动。

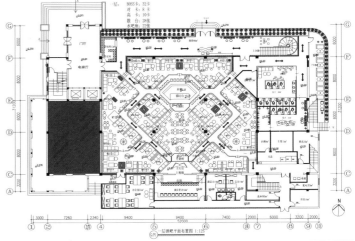

解构主义的流线形建筑外观，极简主义内核，配搭表现主义的梦幻迷离视觉营造，结合全球最尖端的 Magics 和 3D Mapping 立体影像技术，植根于基因深处的"奢华复古与时尚潮流"流韵。力图呈现独特的酒吧品牌魅力，展现独特的酒吧文化氛围，确立在本土酒吧行业中的至尊地位。丰富目标消费群体的娱乐生活，弘扬新时代的阳光酒吧文化。

The first Queen's bar was built in Macau, in 1998 while 2011 saw the first flagship of Club Queen in Hangzhou, then 2012 in Zhengzhou and 2013 in Nantong which starts the mode of an ultimate night club on its authentic brand.

The one in Nantong is aimed to break away the usual building form to revert people's perception and understanding of space to present a future construction seemingly floating above water. The pioneering shape that the volume looks away from the gravity achieves a mirage effect. The tipple of acute angle and the silk-flying curve allows for unprecedented visual effect. The space here is flexible to change its body as freely as earth.

Beneath the skillful hands of the designers, the reality is interwoven in abundance while always changing energy is captured. The modelling that is of multi perspectives, mobile fast and combined with high-tech structure make an image rich in connotation. The spatial geometrical structure is boldly used to create novelty beyond the reality, from which urban vigor and vitality is revealed.

Minimal in its core, the streamlined deconstruction with fantastic expressionism and 3D imagine technology of Magics and 3D Mapping is rooted in retro extravagance and fashion. All is to present bar charm in bringing out a unique ambiance in the space for the paramount position of the bar inside the local industry while enriching the entertainment and overspreading a healthy and active bar culture.

广州增城迷笛会酒吧
MINT BAR

项目位置：广东增城
设计公司：深圳市新冶组设计顾问有限公司
设计：陈武
参与设计：张春华
主要材料：密度板雕刻、水磨石、雅典金花大理石、香槟色钛金、氟碳漆、实木地板等
面积：1 338 m²

Location: Zengcheng, Guangdong
Design Company: Shenzhen Newera Design Consulting Co., Ltd.
Designer: Chen Wu
Participate: Zhang Chunhua
Main Materials: MDF Carving, Terrazzo, Marble, Fluorocarbon Paint, Solid Wood Floor
Size: 1,338 m²

夜，是属于光与电的盛宴；
城，是欲望与享乐的载体；
迷迪，是一种人群，也是一种生活；
迷迪之城，是放大了的欲望天地？还是缩小了的迷离都市？
当白天的喧嚣逐渐远去，光与电的盛宴开始成为这座城市的主角。

脱去朝九晚五的刻板，每个人不再是商业机器上的环节，伴随着自我的回归，每个人都渴望拥有自己的小时代，把握属于自己的话筒。

迷迪会就是这样的话筒所在，在石材外墙的衬托下，这个带有20世纪工业风格的建筑显得颇为宏伟，而与此形成鲜明对比的是建筑中多达20个的方形玻璃窗，紫色、黄色、绿色、蓝色……如同调色板一样的拼花玻璃窗背后，是怎样的世界？

终于找到了出口，当我们将视觉聚焦到建筑中间的方形门框内，门厅墙上一串深紫色几何图形在留白的空间中营造出了某种特定的意境，是叶子？还是水泡？还是花瓣？让人怎不想一探究竟？看似朴实的工业时代建筑皮肤下，蕴含着时尚的生命活力，岂不正是广州增城这座工业新城的写照？

最显眼的，"还是MiNt CLUB"中的"i"，它代表着话筒，也代表着自我，也折射着夜店文化的变化，绚烂、刺激的环境逐渐让位于个人互动和需求。

踏进大门，仿佛置身透明的水底溶洞。周围层叠的不规则圆形几何镂空墙板仿佛围绕着溶洞的水波，流动而又随意。这样的风格一直延续到一楼的舞台大厅，围绕着T型舞台和DJ台，大厅中的散台与卡座依次排开。

但是更吸引人的是充斥着整个空间的不规则几何图形，在光影的营造下，既满足了空间造型的需要，也满足了功能的需要。从背景灯，到LOGO墙，到酒桌，到扶手，是与现有秩序的冲突？还是对我们某个梦境的实现？也许，这就是属于声光电世界的，真实而又不真实。

与造型相比，设计师更加注重的是科技互动在空间气氛中的参与，当音乐与光影同步，在斑驳的片段中，每一个散台或者卡座，都成为了属于你的空间。

属于夜的文化是多样的，这也许就是夜之所以吸引我们的所在，在大厅的激情与绚烂之外，二楼的包房则要冷静许多，属于金色与玻璃的专属冷峻色调逐渐取代了楼下颇为活跃而五彩的背景色，金属的气泡造型在楼梯拐角处不着痕迹地完成了这个过渡。

与大厅服务模式不一样，在包房的设计中，低调、华丽又有内涵的空间颇有进入殿堂的感受，这个空间专为年龄层次和消费能力更高的夜店消费人群打造，不管是沙发音乐，还是明信片式的包厢气氛营造，亦或是随处可见的金属色调，在这里，光影的效果完全让位于空间使用者的需求，只有墙面上偶尔出现的金属水滴空间造型元素和石材扇形造型元素，还依然延续着这座迷迪之城的后现代特征。

一座城市，其实是不同圈层的聚合，他们有不同的需求，也有不同的娱乐方式。设计师为迷迪会所注入的要点，并不在于技术或者造型层面，而是更多地关注到空间使用者与空间互动的层面。迷迪之城，其实是人与欲望的互动之城。

Night is only for the feast of light and power.

City is a carrier to hold desire and enjoyment.

Mint Club completes the desire and the life of people.

And does this project makes an exaggerated lust paradise or a shrunk blurred metropolis?

As the bustling and hustling dies way from daytime, light and power are beginning to come onto the stage of night.

Everyone, a chain on the mass production, is realizing its return to ego longing for an era of their own to have its own voice tube. Here presents such a voice tube: against the stone wall, the building with obvious character of the industrial 20th century is set off with the more 20 square glass windows, whose hues of purple, yellow, green and blue seem to have made into a pallet. What on earth is a world coming?

The square door frame in the volume central is so eye-catching. The geometric purple pattern on

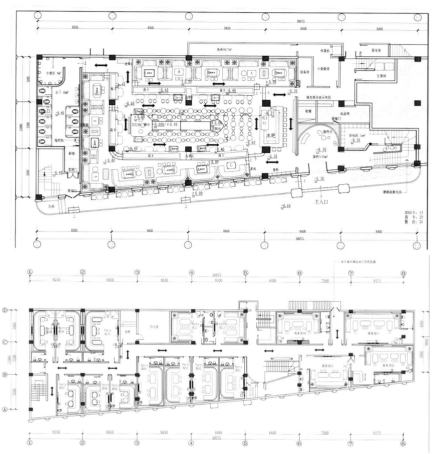

the foyer wall with its surroundings left blank presents some specific concept image. You cannot help but wonder whether it's like leaf, blisters, or petal. Such a curiosity ignites you to explore further. Doesn't the vigor and vitality behind the simply industrial skin make a portrayal of the newly-prospering industry city of Guangzhou?

What keeps the most conspicuous is the "i" letter of MiNt Club, which represents the voice tube and the ego, while reflecting change and brilliance of the culture of a night entertainment avenue. It's here that the provocative environment gradually makes room for personal interaction and individual meet.

Once within, you seem to be exposed in a karst cave in water, where the irregular circles overlapped around, hollowed out wall, are life ripple, flowing free. Such a style is continued into the dance ball of the 1st floor. Around the T-stage and DJ, are seats and docks. The irregular geometric pattern against the light and shadow is not aesthetic and functional. Do items like backdrop light, LOGO wall, table and railing make conflict with the orderly or realize our dream? Perhaps that's exclusive to the world of light and electronic, real and virtual.

The interaction on shaping the ambiance by technology is laid on more emphasis. Light and shadow dance to the music. Each seat and every deck is just for you.

Culture for night is diverse. That's what it attracts us. Compared with the lobby, boxes on the 2nd floor is much calmer, where cold hues of gold and glass takes place of the bright colors downstairs. And the metal bubble around the stair corner completes the transition.

Unlike the lobby, boxes are designed low-key and luxurious. Palace it is like. Here takes as its targets people older and with more consuming power. Whether sofa music, the postcard-like box, or the metallic everywhere spares no effort to accomplish such a mission. The effect of light and shadow give way to the needs of people within. The occasion metal water drop and the stone fan continues the post-modern feature of the city.

A city actually provides places for people to gather around, who have various needs and different entertainment. The design merits of this project lie not the technology or modeling involved, but the interaction between people and space. A space it's really is of desire and interaction.

流光溢彩 灯光设计
THE COLORS LIGHTING DESIGN

安徽芜湖星光璀璨娱乐城
SPARKLE ENTERTAINMENT CITY (WUHU, ANHUI)

项目位置：安徽芜湖
设计公司：汤物臣·肯文设计事务所
设计师：谢英凯
主要材料：蝴蝶米黄石、银镜、灰茶镜、木纹防火板、不锈钢、羊毛地毯

Location: Wuhu, Anhui
Design Company: Inspiration Studio
Designer: Thomas Tse
Main Materials: Marble, Mirror, Anti-Fire Board, Stainless Steel, Wool Carpet

项目位于芜湖中心商圈，周边多个大型的商务中心及大型酒店，拥有得天独厚的商业环境和消费市场。本案以国内高标准的娱乐场所规范进行设计，吸引不同年龄不同阶层的人士玩乐、消费，为本地客人及周边城市客人提供服务，营造健康休闲娱乐氛围，活跃芜湖文化产业。

本案项目是当地一个大型综合娱乐城，综合了精品酒店、大型演艺中心、专属私人会所等项目。设计师通过对空间及人流动线的匠心独具的设计，让每个空间既关联又独立，每个区域更设专属入口，令来到这里的人都能轻易地找到属于自己的享乐空间。

精简而舒适的商务空间是本案酒店设计所追求的。在风格上，设计师采用了简约、精致的现代风格带出空间的时尚气派，满足了精英人士对品质生活的追求，极具个性的家具摆饰更突显空间的灵动性，种种精心的布局设计体现出设计师以人为本的设计理念，客人身在其中，消除了工作带来的压力与疲倦，使身体和心灵都得到放松。

大型演艺中心在设计时即定位为大众化娱乐区域。设计师通过众多娱乐元素的穿插与组合，使空间灵动且多变，让空间的语言变得丰富多彩，令不同的房间呈现迥异的风貌，使不同层次的消费者能根据自身需要选择不同的空间。通过对大众化娱乐的带动，本案将成为芜湖时尚的风向标，以娱乐带动当地经济的繁荣。

专属会所的设计为商务洽谈区域。讲究疏、漏、透的空间层次结构，令到室内各个空间的人相互关照，彼此呼应，步移景换，处处有不同的景致，让整个空间布局浑然一体，拉近洽谈者彼此之间的距离。而错落于各处的商务书房、红酒吧、雪茄吧等不同区间，提供了私人专属的独享空间。

This project is located in center of business district in Wuhu City and surrounded by multiple large commercial centers and hotels, possessing a favorable commercial environment and consumer market. Designed under the high standard of national entertainment public place, this project takes as its target the people of different ages and classes with services for the local guests and the people from neighboring cities as well. It presents a healthy entertainment setting and brings vitality to Wuhu's cultural industry.

This entertainment complex combines splendid hotel, large Performance Center, exclusive private club and so on. Through the tactful design in line with the space and flow of people, all the rooms are interconnected and independent as well. And every zone has its exclusive entrance which provides convenience for the guests to find the specific room to relax.

In this hotel project, we aim to give delicate and comfortable business space for guests. The concise and delicate modern style brings about fashionable charm to space, meeting the elites' pursuit for high-qualified life. The decoration of special furniture which is unique to show the felicity of space. All elaborate layout embodies the people-oriented design concept, allowing the guests

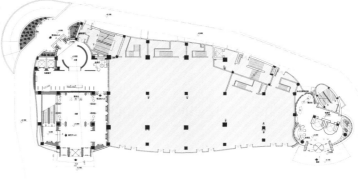

室内灯光 KTV 会所
INTERIOR LIGHTING KTV CLUB

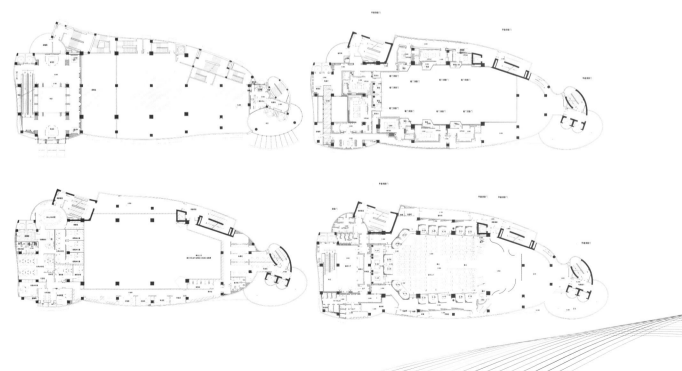

to remove the pressure and weariness after a day's hard work and relax physically and spiritually.

The large Performance Center is constructed as a popular entertainment place. Numerous entertainment elements are interleaved and set together to construct a flexible and changeable space in the center. Various styles and features are thus presented in different rooms fit for consumers from different classes. With the popular entertainment place, this project leads the fashion of Wuhu and flourishes the local economy through entertainment.

The private club is positioned as the place for business negotiation. Focusing on the scanty and penetrating spatial layout, every room of this zone is interconnected and the scenery changes with their every step in this integral space which makes the businessmen feel closer in the negotiation. Besides, the business study, wine bar, cigar zone and the like are scattered about the center, providing some private zone for guests.

流光溢彩 灯光设计
THE COLORS LIGHTING DESIGN

拉科瓦舞场会所
LA COVA DANCE THEATRE

项目位置：西班牙巴塞罗那
摄影师：拉斐尔
面积：500 m²

Location: Barcelona, Spain
Photographer: Rafael Vargas
Area: 500 m²

"拉科瓦舞蹈剧场"，其实是面积达500平方米的夜总会，位于巴塞罗那附近。空间以照明、帷幕、壁画、背光，共同打造轻松之气氛。与众不同的外表、自然的迷人气质，以其如同城市"冰洞"般的气场，为人们提供一个交友、聚会的极好去处。

流动的内里空间，有着很多不同的分区。但大部分开放态势，环形支线，有利于人们互动。前面的公共区域，极尽透明玻璃用材，进步强化了这种易于流动的态势。DJ及观众区位于后部。舞台位于中央。最受到限制的空间，如更衣室、仓储室，恰当地位于空间的最深外。

光照堪称特色。两个新巴洛克式的巨灯，如同钟乳石一样，悬于洞顶。一些站立的花瓶，如同破土而出的石块，但实际却是别样形式的光源。白色的面体，一旦有了不同光照，便有了放大光华、改变空间属性的效果。方便着似乎正置身于洞穴中的人们，让他们去探索最富有魔力的放松场所。家具铺陈全部出自一位名师之手。醒目的酒红色沙发，以新巴洛克式风格重饰经典之风，但同时为空间增加了一抹原生的创意。沙发对面，一组相互咬合的软躺椅完全应业主的要求而设。如同伏地爬行的蛇，那完全是一种有机的形式。

不同的区域，高度不同。因此，除了舞台，各个空间陡然抬升，如同大自然中的地块凹凸不平。

PLANTA. LA COVA TEATRO DANCE

La Cova Dance Theatre, located in Mataro, near Barcelona, is a nightclub with a total area of 500 square meters. La Cova Dance Theatre is a space replete with the lighting, curtains, murals and backlighting to create playful effects. It has a distinct appearance and a natural charm that is ideal for meeting people, and where better to meet a special someone than somewhere as special as an urban ice-cavern.

The interior is a flowing space with different areas, most of which are open and easy to move around in, creating circuits in which people can interact. To emphasise this ease of movement within the public areas, the frontage is made of transparent glass. The DJ cabin and the area for the audience are at the rear, while the stage is in the centre. The most restricted areas, the changing rooms and storerooms, are in the innermost recesses of the premises.

The lighting is one of the main features, with the dancefloor illuminated by two enormous neo-baroque lamps hanging own like stalactites. The other source of light is compensated by some standing vases that look like rocks emerging from beneath the floor. This all white surface has the effect of multiplying the illumination and altering the overall mood of the space by changing with the different colors while the clarity is sufficient to glimpse the people occupying the cavern and to discover the most magical recesses. The decoration is completed by the furniture designed by Elia Felices Interiorismo. There is a striking wine-colored sofa with neo-baroque forms that reinterprets the classical style and adds a dash of originality. A series of interlocking pouffes occupy the opposite side, distributed in accordance with the user's requirements. The organic forms resemble a serpent slithering over the floor.

The different areas are also distinguished by different heights so that, apart from the stage, there are also raised areas along the sides, recreating the irregular terrain found in nature.

180 流光溢彩 灯光设计
THE COLORS LIGHTING DESIGN

室内灯光　KTV　会所
INTERIOR LIGHTING KTV CLUB

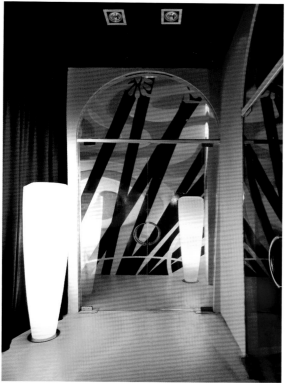
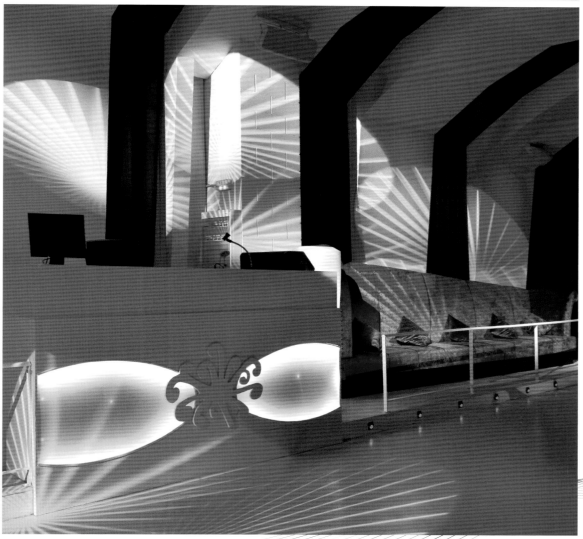

流光溢彩 灯光设计
THE COLORS LIGHTING DESIGN

以色列啤酒俱乐部
"FORUM CLUB"-BEER SHEVA-ISRAEL

设计公司：NP建筑有限公司
设计师：尼尔、韦斯
摄影师：盖

Design Company: Nir Portal Architects Ltd.
Designer: Nir Portal, Shahar Weiss
Photographer: Guy Franko

　　25年的经营使其成为以色列沙漠边缘最有传奇色彩的酒店。25年的风雨之后，2 000平方米的空间以其不同的魅力吸引着来自市区的年轻人。即便1个半小时的舟车劳顿，只为空间里即将上演的PARTY。

　　廊台从1楼升至2楼，为空间创造了一种直线性的感觉。在这里不仅可以举行派对，还可以举行音乐会。公众、贵宾、演职人员各有其空间。所有扶栏都镶嵌有LED照明，平面的布局意象特别强烈。定制的卤素台灯对比性地洒下一地温暖的光芒。

　　16个酒吧空间，扶栏与照明一脉相承。每个酒吧上层空间的酒架全为钢网、金属框架。主俱乐部墙体装饰着彩纸的定向刨花板。板面图案饱含着工业化的明亮气息，却也显得那么温暖。射灯的光束照于墙面，空间的界限泾渭分明。

　　天花空间有着600个LED灯管，每一个灯管的位置都在电脑上标注。天花成了电脑的巨大屏幕。5米高的动感音响屏幕把整个空间分为4个单独的小俱乐部，各有其DJ及音乐播放。

　　除了主厅，整个空间还有很多隐秘的私有空间，这是贵宾们的天地。小小的情侣包厢，躲避着众人的眼。静静的角落里，只有你，只有我。

After 25 years of running business and becoming the most legendary club in the desert periphery of Israel, the "forum" had to relocate to new space and to invent itself from the beginning. Spread over his 2,000 square meters, the new forum, have become the reason Youngers are driving 1.5 hours from center of Israel to participate in its parties.

In order to create hierarchy spaces we have created tribunes rising from ground floor to 1st floor. Thus tribunes serve both party and concert show; it separates crowds and enables unhidden views during thus events, and preform perfectly in case of siting concert. All railing of tribunes and gallery has led lighting that project thru the glass, creating strong image of layout. A custom made halogen table lamps creates contrast warm lighting.

In club area we planted 16 alcohol bars, all lighted in same way as railing. Above each bar hanged an alcohol display made of steel net and metal frame. Major club walls covered with OSB printed panels. Graphics on panels were designed to bring atmosphere of industrial yet warm atmosphere, and thus walls were washed with narrow beam spot light, to

create dramatic boundaries to club.

600 led tube are hanged from ceiling, each tube has its own IP address, so all ceiling can act as huge screen controlled by light jokey. 5 meter high moving acoustic screen can divide the space into 4 separate small clubs; each can have its own DJ and music.

Except of the main hall, the club has a lot of hidden and more intimate areas, for different kind of hang out. Little love nests, hidden from eyes, is spread all around so lovers can find a quiet corner just to their selves.

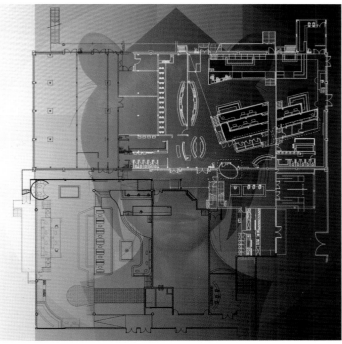
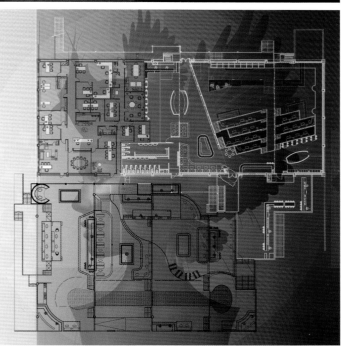
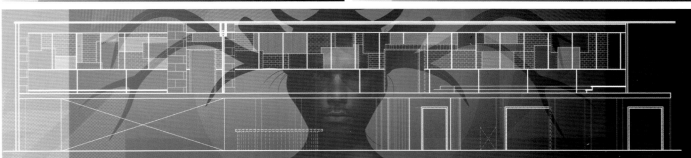

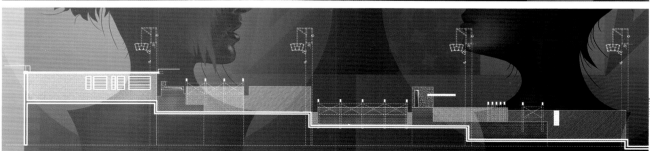

B 流光溢彩 灯光设计
THE COLORS LIGHTING DESIGN

大同 LOVE 100 酒吧
LOVE 100 CLUB

项目位置：山西大同
设计公司：深圳市新冶组设计顾问有限公司
主案设计：陈武
参与设计：吴家煌、代浩
主要材料：古木纹、亚马逊大理石（亮面）、（玲珑玉）透光石、热带雨林、铝板、打砂黑钛金、艺术漆（深灰色）、金属漆、硬包皮革
面积：763 m²

Location: Datong, Shanxi
Design Company: Shenzhen Newera Design Consulting Co., Ltd.
Chief Designer: Chen Wu
Participate: Wu Jiahuang, Dai Hao
Main Materials: Marble, Veneer, Aluminum Board, Paint, Leather
Area: 763 m²

　　LOVE100酒吧是新冶组在原唐会酒吧基础上的全新升级。经营面积1 200平方米。新冶组设计以多年实践经验和锐意创新，结合当地本土文化，打破分解既存的陈旧空间形式、格局和模式。以最新潮的娱乐文化理念，演绎激情张扬的空间情调。时尚动感的节奏，融视听之享受，更好地迎合当下大同娱乐消费市场。

　　风格主张新旧融合、兼容并蓄，整体偏于新派电子风，同时不失中庸之道，与大同这座古城遥相呼应。大厅矩阵灯光应用演绎先锋概念的灯光艺术，圆形场布局带来的是包围式的怡然放松。设计师以曲线和非对称线条为最小单位，在设计细节中把玩非理性因素带来的反叛，刺激和调侃。花叶等自然意向在墙面、栏杆、窗棂和家具上的装饰应用，赋予无机的世界有机的情调。整个空间立体形式都与有条不紊、有节奏的曲线融为一体。

　　360度全方位立体化、激情热舞现场，将国际尖端高科技与R&B、Hip-Hop文化多重混合，在科技电子和节奏舞曲中的缝隙空间，寻找最深入的摇晃点位，不单单是华丽的外壳，这就是LOVE100的个性，充满激情和热情，冷静和稳重。

LOVE 100 is a project transformed on the basis of two bars with a business area of over 1,200 square meters. Experience accumulated over years and innovative creation of the designers is combined with the local culture, breaking away the old spatial organization, structure and pattern. The latest recreational culture is turned to for interpreting the enthusiastic property in the space. The fusion of the dynamic and the audio-visual enjoyment well caters for the local entertainment trend.

Eccentric in style, the holistic style integrating the old and the new is mostly electric. The all-embracing echoes with the old city of Datong. The matrix lighting makes the pioneering light art while the circle avenue brings forward pleasure and comfort. Curves and asymmetrical lines are available hither and thither. Out of elements of non-rationality are some against orthodox, stimulation and excitement. Floral leaves are applied onto walls, rails, window lattice and furniture, the inorganic world thereby getting more sentiment. The whole solid form is therefore blended into the orderly and rhythmic curves.

360-degree 3D design overspreads passion everywhere, mixing the state-

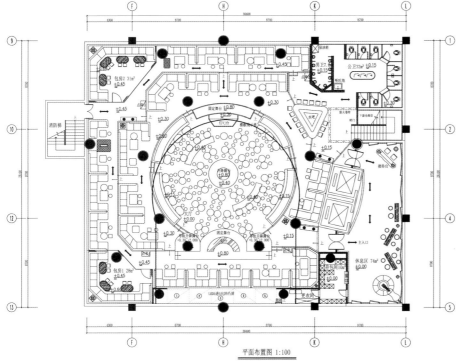

平面布置图 1:100

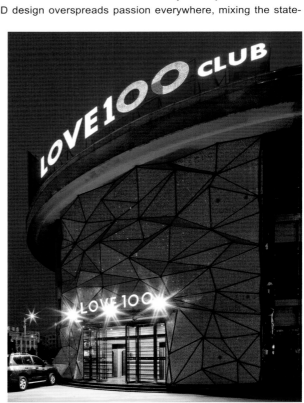

of-art science and technology with culture of R & B, and Hip-Hop. The deepest shaking point location is natural to find in gaps of science and technology, electronic and dance music. Not just extravagant or magnificent in appearance, LOVE100 has its own character. That is passion and enthusiasm, calm and staid.

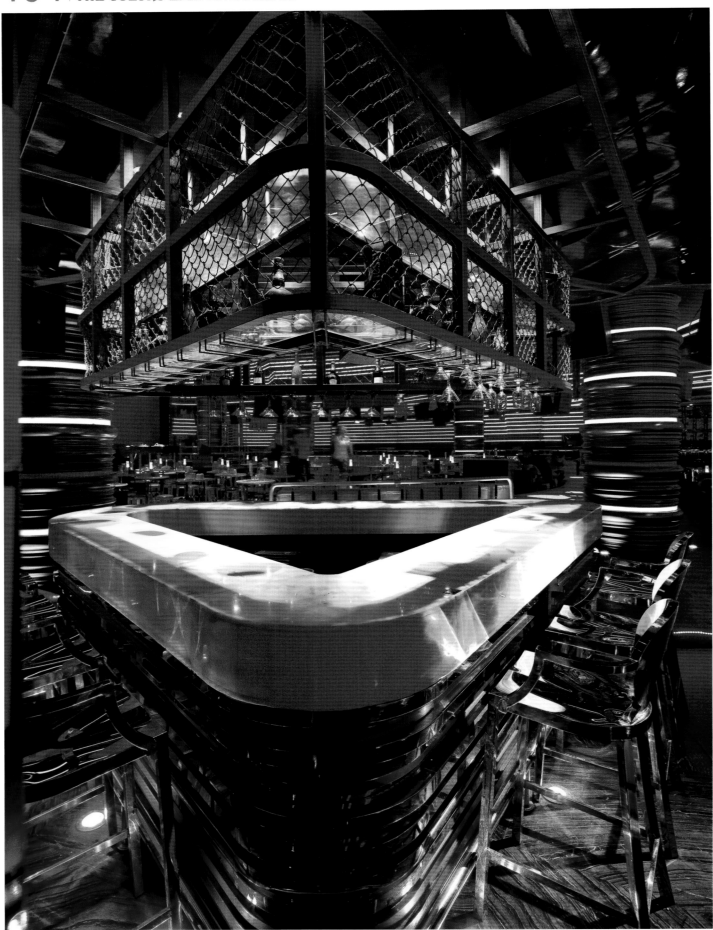

流光溢彩 灯光设计
THE COLORS LIGHTING DESIGN

重庆 CLUB ONE 酒吧
CLUB ONE, CHONGQING

设计公司：深圳市新冶组设计顾问有限公司
设计师：吴家煌
参与设计：张春华、王松涛、代浩
主要材料：斑马石、卡里冰玉、云石马赛克、大花白、啡金、水晶马赛克、肌理漆、黑钛金
面积：1 500 m²

Design Company: Shenzhen Newera Design Consulting Co., Ltd
Designer: Wu Jiahuang
Participate: Zhang Chunhua, Wang Songtao, Dai Hao
Main Materials: Marble, Mosaic, Texture Paint
Area: 1,500 m²

　　重庆 CLUB ONE 派对酒吧位于重庆的酒吧聚集区——解放碑得意世界。"得意世界"是重庆夜店的标志，也是重庆的潮流聚集中心。小小一方天地，集中了三十多家大小不一的娱乐场所。"得意世界"将夜生活颠覆成了一种个性的享受。

　　CLUB ONE 是重庆首家以"轰趴派对"模式打造的高端酒吧。作为顶级奢华为理念打造的 CLUB ONE，耗资 1 500 万精心打造，彰显顶级奢华的酒吧娱乐新体验。推出全新的娱乐模式，汇集韩国、日本等亚洲国家顶级娱乐元素，为重庆带来"国际化"、"明星化"、"派对化"、"高端化"的酒吧娱乐新体验。本着"国际、创意、格调、质量"的品牌延伸概念，设计师以通透的空间架构极大地提升酒吧空间视听效果，前卫的设计增强对感官的直接冲击力，让身处其中的朋友充分地释放情感，在夜晚得到彻底的放松。

　　在 CLUB ONE 的设计中，干净利落而富于设计感的直线元素贯穿始终，纵横交错，无限蔓延的线条，给观众以震撼的视觉体验和前所未有的审美感受。对点、线、面的艺术处理，是空间艺术设计中最基础的元素。点、线、面的巧妙组合与穿插应用构建了艺术作品的创作基调。不同形式的直线条构成了 CLUB ONE 延伸多变富有动感的视觉效果，在灯光的映衬下营造出多层次的虚空间。解构切割重组的设计手法，创造出丰富多彩的环境氛围。在空间整体色系上，设计师以高雅的冷灰搭配温暖的黄色，呈现时尚与经典的兼容并蓄，达到一种简洁、舒适的氛围。

　　在酒吧外观的设计上设计师大胆采用倾斜、交错、叠加等效果勾勒出新派酒吧的先锋时尚。前卫的造型搭配明亮温暖的色系，不动声色的奢华与张扬，让观者眼前一亮。与 CLUB ONE "国际化" "高端化" 的品牌定位完美契合。

　　为了强化空间的直线视觉效果，大厅吊顶部分的开放式设计十分新颖。一来化解顶部厚重感，提升了顶部视觉水平线。纵横捭阖，多变化灵动之美，以强而有力的穿透感缓解室内空间压抑；再者，改变了墙面单一的视觉感，交错的线条由天花一路延伸至墙壁，形成连贯一致的效果，保持了整体设计的统一性，更是进一步强化了直线条的空间特点；通透的设计与选材更利于大厅的声音传播。

　　包房采用水平对立设计，整体设计贴合空间特点，在确保满足使用功能的前提下达到审美功能的最大化。凹凸式吊顶构造复杂而富于变化、层次感强。电视背景墙选择层次丰富的线板堆栈，呈现层层变化的精致度。墙头圆弧造型，搭以转角处线条延伸，在视觉上呈现立体感受。

　　在后期配饰上，也多选用后现代风格灯具饰品做室内渲染特点，提升整体效果。此外没加入过多零碎的装饰，原始的创意与细节却经得起时间的磨砺。动与静，暧昧与陌生，微妙非凡。

A project Club One is that is located in Complacent World, a landmark for the night entertainment avenue in Chongqing. An area it is, though small, clustered with over 30 casinos. And Complacent World subverts the night life into an individual relish.

As the first high-end pub modelled with idea of Home Party, Club One treated with top luxury is accomplished at the cost of 15 million RMB to embody a completely-new bar and recreational experience, where to converge leading amusement elements from Asian countries to offer Chongqing new visions of Globalization and Orientation of Celebrity, Party and High End. With brand concept "internationality, novelty, style, and quality" centered, the transparent spatial structure maximizes the audio-visual effect, where the latest design boasts the impact exerted on the organ senses, so that people within are able to release their emotions to a large extent for head-toe relaxation.

Direct lines are interposed and interwoven vertically and horizontally throughout the space. When going limitless, lines present a shocking visual experience and an unprecedented aesthetics. The collection of points, lines and surfaces makes the most basic and fundamental element, when pieces of art work are involved concurrently. The numerous line forms allows for visual effect that is changeable, dynamic and extendable. Against lighting is the virtually multi layers. Design approaches of deconstruction, cutting and reorganization bring about rich and colorful ambience. As for the holistic tone, the cold gray coordinates with the warm yellow to appear both modern and classical, eventually creating a concise and comfortable atmosphere.

The building appearance boldly employs stilting, interweaving and accumulating to outline the pioneering and fashion of an up-to-date club. The pioneering modeling and the bright, warm hues are of luxury in a calm yet equally flamboyant manner for views to feel enlightened all of a sudden, impeccably consistent with the brand position of top level.

In order to strengthen the vision effect by direct lines, the open suspended ceiling in the lobby is of clear and artful finesse, offsetting the heavy sense and raising the visual horizon to provide lithe beauty. Meanwhile, the simplex vision of the wall is altered when lines spread from the ceiling to the wall, making a consistent effect in keeping the unity and further intensifying the vertical features. The design as clear as crystal and the

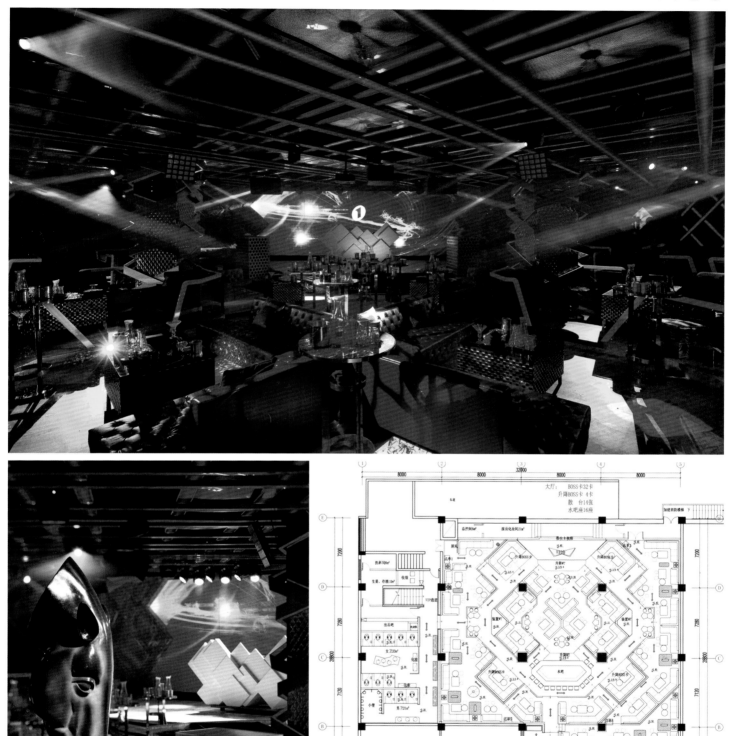

selection of material are reciprocal to the sound transmission.
The boxes are aclinic and antagonistic. In accord with the whole spatial character, the design maximizes the aesthetics on the premises functions are met. The concave-convex suspected ceiling is complicated and of rich layers. The backdrop for the TV is of line board to make delicacy that changes with each layer. The wall top is arc with nook extended into lines to offer stereoscopic feelings.

In order to render the internal, accessories and furnishings are embellished with post-modern lighting fixture. Barren of surplus embellishing trifles, the primitive creation and details are bound to be able to take the test of time. Dynamic and statics, and ambiguous yet unfamiliar, all are subtle yet marvelous.

B 流光溢彩 灯光设计
THE COLORS LIGHTING DESIGN

台北纯 K 杭州店
CHUN K PARTY (HANGZHOU)

设计公司：深圳市新冶组设计顾问有限公司	Design Company: Shenzhen Newera Design Consulting Co., Ltd
设计师：陈武	Designer: Chen Wu
参与设计：吴家煌、代浩、张春华	Participate: Wu Jiahuang, Dai Hao, Zhang Chunhua
主要材料：云石马赛克、大理石、钛金、实木、墙纸、工艺玻璃、玻璃膜、水泥板型板	Main Materials: Mosaic, Marble, Titanium, Solid Wood,
面积：2 014 m²	Area: 2,014 m²

在台北纯K杭州店的创作过程中，设计师充分考虑到了杭州冠绝江南的士大夫情愫，他们将江南园林艺术与空间设计融合起来，通过对罗马柱、工艺玻璃、窗格、立体电话亭以及巧妙的空间隔断的利用，营造出移步换景、景中又有景的奇妙效果。不论是转角处的红色沙发空间，还是街边电话亭，抑或是包房中的星空，不同特色的空间交织在一个平面之中，又相互渗透。

走进纯K Party的大厅，玫瑰金材质、玻璃以及LED灯似缎带亦似流水般地将空间完整地结合起来，水波光墙及水晶般的灯效如水波浪潮般一层一层流动着，给人一种温暖、尊贵的感觉。在主色调的选择上，纯K Party摒弃了传统炫目的黄金色，选用色调柔和、瑰丽迷人的玫瑰金，更好地体现了金属材质的精致和细腻。灯光绚丽的效果可通过科技和设计手法去实现，但若没有丰厚的文化内涵，空间也就无法给人深刻隽永之感。因此，台北纯K Party的灯光设计力求与玫瑰金属材料巧妙结合，不沉迷于金银饰的奢华风格，同时也丢弃了浮华的心理，更多地追求音乐和视觉效果，时尚娱乐的风格，迎合消费群体不断提升的消费品位。空间全新的动感灯光设计，可进行多种模式调节，营造出多层次的视觉效果。

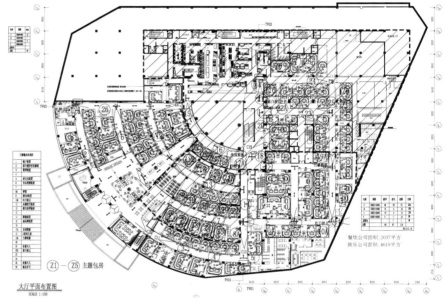

大厅平面布置图
SCALE 1:150

此外，设计师更把"疏影横斜水清浅，暗香浮动月黄昏"的美丽意境延续到空间中，或柔媚如影院，或明亮如大堂，或昏暗如水吧灯光，对空间做出了不着痕迹的划分，让客人们有着更适合的体验。

公共空间以中轴线为基础，让客人能够以最快的速度进入自己想要去的目的地：影院、水吧、礼品屋或是已预定好的包房。不同的包房区域又采用对开门做遮挡，相对私密的空间，让客人们更加舒适。包房设计力求艺术性与功能性的完美结合，别具特色，拒绝雷同，统一采用吸音的软质材料表现音响混响效果，更具娱乐氛围。包房中还特别增加自助式酒吧区、小型舞池区、情侣品茶区、小型舞台表演区等自由娱乐区域，这也是对KTV附属功能的巧妙补充。包房中的立体沙发、吧椅设计遵循人体工程学原理，让客人坐卧舒适，轻松愉悦。

作为全新K歌潮流创领者，台北纯K Party的装潢设计不仅重视对美学、技术和经济的把握，更将人性、人文提升到了一个高度。其装潢设计围绕客人的舒适、放松体验展开，在格局、装修上通过对细节、选材的雕琢，创造性地解构人文素材，实现功能性与艺术性的统一，唤起人内心深处对美的向往和记忆。

金碧辉煌的大厅，低调而奢华的包房，由云石马赛克、大理石、钛金、实木、墙纸、工艺玻璃、水泥板型板构成，在简欧风格的家具、水晶灯、富有欧洲风格五彩油画的妆点之下，恍如欧洲华庭。仔细观察可发现，空间随处可见云纹的帷幕、前台背后的中国风窗格、水吧窗口的墙垛风横梁，中国风十足。

The process of this project takes account of the local emotion for literati and officialdom in Feudal China by getting one Gardening of the low-reaches of Yangtze River and the space design: the finesse and ingenuity to use marble pillar, craft glass, window of lattice, 3D telephone booth and the skillful partition makes a marvelous effect that scenes change with step going, and another one is embraced with a scene. Sections with distinguishable features are under the same roof, intersperse and interwoven, like the red sofa in the nook, the telephone booth, and the starry sky in boxes.

The hall is fused in chorus with rose golden material, glass and LED belt and strip. The wavy wall and the crystal lamp seem to be flowing, generating feelings of warmth and dignity. Rose golden poses to be the main hue instead of the usual golden: softer and more intruding to better embody the delicacy and subtly of the metal material. Lighting and rose golden material are combined to render a profound culture backdrop for the lighting brilliance. The space is barren of luxurious style of gold and silver embellishment as well as showy mental needs; on the contrary the space is emphatic that music, visual effect and style of fashion and entertainment is pursed to cater for the taste of the consumers that has been upgraded. The completely-new lighting is adjustable in a multi mood to generate vision effect of more layers.

Additionally, the artistic concept is implanted within that space shadows are reflected horizontal or slanted into the shallow water while fragrance

comes secret with the moon reveal its veil as night comes. So sections allow for different experience for people with different demands, the cinema as marvelous as possible, the hall maximized bright, and the water bar dim as unexpected.

The public space is centered on the axis, along which guests can enter their destination at a fastest speed, like the cinema, the water bar, the boutique room and the private rooms. Boxes are endowed with a double door to ensure relative better privacy and more intimacy. All boxes are destined to combine the artistry and the function while sticking to be kept individual and equipped with acoustic material to boast the recreational air. What's more, self-entertainment areas are fixed, like self-help drinking, small-sized dance floor, tea drinking for lovers and small stage. A real supplement this is for KTV functions. Sofa and chair in the boxes are more comfortable and pleasing when in line with the principles of human engineering.

As a leader inside, the project of K Party not only stresses the grasp of aesthetics, technology and budget, but reaches humanity and culture into a new level. All design and decoration are guest-friendly to maximize ease and comfort. The focus on detail, the selection of material, the use of culture, and the combination of the function and the artistry are evocative to the beauty in the depth of heart.

The hall brilliant and splendid and the boxes reserved yet luxurious are of mosaic, marble, titanium, solid wood, wallpaper and so on. The simple European furnishings, the crystal lamp, and the paintings of European five colors seem to have presented a good dwelling that comes across Europe. Once attentive, you no doubt find that, everywhere are curtains patterned with auspicious cloud, Chinese windows of lattice, and pier and beam fixed above window of the water bar.

206 流光溢彩 灯光设计
THE COLORS LIGHTING DESIGN

B 流光溢彩 灯光设计
THE COLORS LIGHTING DESIGN

皇庭 KTV
ROYAL COURTYARD KTV

计公司：齐合作设计　　　　　　　　　　　　　　　Design: Chiho & Partners

本案以休闲和的士高为主要功能，力求专业、刺激。墙身的有孔金属板与天花板融为一体，同时也达到绝好的吸音效果。强烈的色彩衬出专业的灯光效果和音响空间。包厢内家具简单、个性、舒适。沙发色彩突出个性，超尺寸的设计创造舒适的环境，让消费者感到既有娱乐性又有观赏性、趣味性等。

Royal Courtyard KTV focuses on functions of leisure and disco, when striving to be of specialty and excitement in integrating the perforated metal sleets on the wall and the ceiling to achieve a good effect of absorbing sound. Against the strong hues is set off the lighting and acoustic space. Furnishings in the boxes are simple, unique and comfortable. The sofa color highlights its personality while its super large size is aimed for ease and pleasure, so that people can feel the spatial entertainment, ornamental value and enjoyment.

流光溢彩 灯光设计
THE COLORS LIGHTING DESIGN

维的雀旗舰店
VIDEOTRON'S FLAGSHIP STORE

设计公司：李希德建工
顾问：德绍等
多媒体安装：棱罗高科
总承包：艾伯特建筑

Design Company: Sid Lee Architecture
Consultant: GSMprjct, Dessau, Planifitech
Multimedia Installation: Solotech
General Contractor: Albert Jean Construction

"维的雀"旗舰店位于蒙特利尔繁华中心区，店面设计别致、高级。其建筑、设计由几家公司联袂制作，旨在为大众提供集多媒体、品牌、商务为一体的原生态体验。里面装点的羽饰比较罕见。如同家居般的购物场合，是全新的体验。空间正面、中央尽展图文之美，独具特色。

用于产品展示的空间，耗时不到十月。360度的全方位展示，让人一睹其品牌魅力。负责此处的设计公司，同时负责打点"维的雀"的网站，广告品牌推广。

大胆、富有创意的设计，经过团队的努力最终实现了其终极目标：新式的互动产品展示必定赋予其品牌坚实的竞争力。

Videotron opens its new flagship store, a unique, avant-garde business space in the heart of downtown Montreal. The instore environment is the collaborative work of Sid Lee Architecture, Sid Lee, the Videotron team, and a number of partner suppliers. They are pleased to invite the public to experience their original concept, which combines multimedia, branding, and commercial architecture with rarely seen panache. The space will offer an all-new experience in an environment Videotron customers will feel right at home in. "Like Videotron's brand and Sid Lee's campaigns," said Martin Leblanc, Sid Lee architect and partner, "shopping in the store will feature a graphics-oriented environment putting content front and centre."

In less than 10 months, Sid Lee Architecture and its partners came up with the concept and completed Videotron's crown jewel—the place where it will present its full product range. The store provides a compelling example of the impact and brand consistency of the 360 degree approach used by Sid Lee, which also developed Videotron's website and advertising campaigns and handled its branding work.

Sid Lee masterminded this bold, creative project by bringing an entire team of experts together in the multidisciplinary pursuit of an ultimate goal: bringing the brand to life with new interactive ways of showcasing the products in this highly competitive sector.

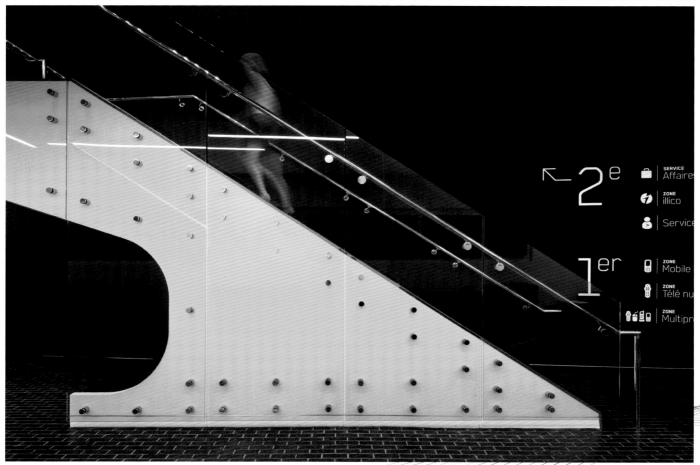

革命酒吧
THE REVOLUTION LOUNGE

流光溢彩 灯光设计
THE COLORS LIGHTING DESIGN

设计师：珍等

Designer: Jean-Franmois Bouchard, Cirque Du Soleil, Stéphanie Cardinal, Nol Van Genuchten, Moment Factory, Virtango, Alain Vinet

"革命酒吧"占地715平方米，内有好几个次生酒吧。中心区，三个支撑性立柱，并以三角形板彰显空间核心。其中一些板材取自发光钢板，另外一些覆以白色隔屏。如此设计灵感，源于《缀满钻石的天空中的露西》。钻石的光华漫射中，立柱却是光源的集中点，对天花起着卫护作用。板面之间，柱体之间，另有系列球点设计。35 000个二色晶状体，悬于金属发光绳之上，如同满天星芒，升华着会所给人的迷离感受。

立柱之下，一字排列着设计于1969年的粉色萨科豆袋椅，并有设计于2005年萨科互动桌台。直线、曲面之间，消抵着隐含于内部的古板。周围其他次生酒吧空间，还有长长弯弯的台坎、沿墙排列的桌台。整体家具的陈设乃是20世纪六七十年代的顶级风。音乐气氛以"4"数字为基础：四个"披头士"乐队，乐队的四个阶段，英语单词Love的四个字母等等。每个舞台，各有其灯光照明，或黑至白、至粉递进。

为了完成对墙面活泼气氛的操控，MF次生酒吧，承袭原先的"窗户"：玄关墙，即位于酒吧之后的18米长墙。另有4个舷窗，是电影《黄色潜水舰》的象征，内置LED系统，通过双面镜材进行影像投影，同时于酒吧上空的下悬式天花还另设有镜材。如此不同的多媒体平台对空间12个不同的形象表面提供着有力的支撑。该理念灵感源于20世纪60年代和风靡一时的"披头士"乐队。

本案主体，"革命酒吧"配有7个低矮、但却可以互动的桌台，任由老年客人"涂鸦"。夜晚时分，称为"执政官"的员工，穿行于客人中，通过随身携带的便携式设施，让客人把其创作投影于中央立柱的屏风之上。

如此浩大工程，历时两年时间，经几家知名公司联袂设计，本案必定于蒙特利尔为世人呈现一个天才之作。它是数量的表现，也是质量的展现。

The revolution lounge itself measures 715 square metres (7,700 square feet) and is divided into several smaller lounges. At the centre, the architect Stéphanie Cardinal used the three support pillars to create a central point of attraction. Inspired by the song Lucy in the Sky with Diamonds, she surrounded the pillars with triangular panels, some of which are made of gleaming steel, while others are white screens. With this covering, she has transformed each pillar into the point of a diamond frozen as it shatters, dragging the ceiling with it in its fall. Inside these structures, lighting designer Nol Van Genuchten has installed a string of spots. To add to the club's psychedelic climate, he has surrounded the columns with a sparkling cascade of more than 35,000 dichroic crystals suspended on metallic chains attached to tracks.

Under the pillars, straight-lined and pink Sacco (1969 design) banquettes alternate with white, interactive tables (2005 design). The contrast between straight lines and curves calls to mind resistance to authority and diktats. The surrounding lounges offer numerous pleasures: long, winding banquettes line the walls and also come with interactive tables. The overall look of the furniture borrows from the pop style of the 60s and 70s. Created by DJ Alain Vinet, the musical atmosphere is based on the number four (a symbolic number: four Beatles, four periods in their careers, four letters in the word LOVE, etc.). Each of four stages features a distinctive lighting display, gradually moving from black and white to hot pink.

In charge of wall animation, the Moment Factory inherited three "windows": the entrance wall, an 18-metre long section of wall behind the bar. To add to the mystery, the Moment Factory created four portholes (symbolizing the Yellow Submarine) equipped with LED systems that project images using a double-mirror system. Finally, the lack of depth imposed the installation of an overhead projection system, using a mirror attached to a drop ceiling above the bar. These different multimedia platforms support 12 different graphic animations in a style inspired by the 60s and The Beatles.

The revolution lounge is equipped with seven low, interactive tables that allow patrons to create their own graffiti. Throughout the evening, staff members called "Consuls" wander among the patrons carrying rings that allow them to upload certain creations onto the "screens" on the central pillars.

It has taken two years to complete this large-scale project. By entrusting Cirque du Soleil with the design of the lounge, INK (Toronto-based project management firm) ensured strong links with the show LOVE. This project is further proof of the quantity and quality of the talent coming out of Montreal.

B 流光溢彩 灯光设计
THE COLORS LIGHTING DESIGN

"八角"会所
CLUB OCTAGON

设计公司：鄂尔坦娜有限公司
摄影师：孙南国
地点：首尔
面积：2 600 m²

Design Company: Urbantainer Co., Ltd
Photographer: Namgooong Sun
Location: Seoul
Area: 2,600 m²

"八角"会所乃为老旧建筑整改项目。应客人要求，旨在于2 640平方米的空间，打造集高科技礼堂、会所、酒吧、饭厅于一体的场所，为客人提供一个音乐的体验。本案一经推出，可谓开创了韩国市场，以娱乐、社交文化统一于同一领域之先河。平面布局、设施铺陈、4D的灯光照明、模块式的座位、甚至VIP室内的冰桶，都如本案名称所示，尽展"八角"之意向。

设计师以拓展各功能空间能量为己任，但却不以引领首尔市场为目标。眼下韩国匮乏仓库、工厂的事实激发了本案设计赋予本案"仓库"、"工厂"般空间的灵感。其中的亮光、高科技元素，在韩国市面上常见。

鉴此，空间对用材加以限制。裸露的钢管、电梯、消防通道、消防设施、通风、氧化水泥地板的使用颇为有限。在满足设计目的的情况下，只为追求弥漫在旧时学校的那一份原始能量。

内里空间，照明、八角结构，尽以展示主题意向。韩国传统中，"八角"元素如同天地的代表，广为人们使用，如皇室所建亭台楼阁、皇家餐具等等。正因为本案空间形状介于圆形与方形之间，"八角"元素适时出现，象征着人类沟通于天地之间。

于是，整个空间、设施、细部尽以"八角"出现，但角度只限于45度、90度、135度。

正是因为"八角"形状的使用，设计才得心应用。于设计看来，圆形的空间原本可以尽展建筑的美学。舞台全方位的观赏角度，适宜各种音箱的安装、调配，声效功能极好。特别是顶级音箱的环绕系统，不管客人在哪里，音效都是绝顶的赞。

Club Octagon answers the client brief of renovating 2,640 over two levels of gutted hotel basement to create a high tech auditorium, club, lounge and restaurant that put music and peoples' experience first. Urbantainer developed a new type of multi-space for entertainment, socializing, and subculture that was lacking in the South Korean market. Conceptually every detail of Club Octagon works with the octagonal form the corporate identity, including the layout, 4D media lighting, modular seating, to even the ice buckets in each of the VIP rooms.

The design cocept for Club Octagon was based on the energy of these spaces, instead of aiming to be Seoul's next trendy club. The lack of converted warehouses and factories as creative multiuse spaces in Korea inspired Urbantainer to create their own whilst including the glossy and technological elements that are a must for the wired Korean market.

With the concept of a factory in mind, the club relies on limited

materials only. The minimal design with elements like exposed steel beams, elevator shaft, fire prevention system, ventilation, and epoxy cement floors to allow focus on the programmatic content of the events, while keeping alive the excitement of the raw energy of old school raves.

The design motif that runs throughout the club interior, lighting, and it sets the structural rules for construction in octagon. Inspired by its prevalence in Korean traditional design (such as in royal pavilions and dishes), Urbantainer chose the octagon as their design motif because of its specific representation of a human between the elements of land and heaven. Since the shape is in between a circle and a square, the octagon resembles a person, the medium between sky and land.

With that as a conceptual starting point, Urbantainer designed the entire space, fixtures, and details following the octagonal motif and setting the design guidelines using only 45, 90, and 135 degree angles throughout the entire space.

The octagonal shape also allowed the designers to create the effect of a coliseum without the design and aesthetic difficulties of building in a completely circular shape. Whilst being optimal for viewing the stage from all angles and producing superb acoustics, the octagon allowed for the presence of straight edges and was better suited for the inclusion of technological equipment into the layout design such as with the Funktion-One surround sound system.

吉隆坡 Rootz 会所
ROOTZ CLUB

流光溢彩 灯光设计
THE COLORS LIGHTING DESIGN

设计公司：精致设计
面积：652 m²

Design Company: Design Spirits Co., Ltd.
Area: 652 m²

Rootz 会所位于马来西亚吉隆坡。值吉隆坡 LOT 10 迎来 20 周年之际，其位于屋顶的停车空间，集俱乐部、剧场、饭店及庭院为一体，款款地走到了大家面前。

独有的气氛，适合世界各地的客人在此舒展心情，翩翩起舞。那是昔时，东西方人士在此起舞的景象。

世界各地的宫殿巍峨的气势自在这里体现。更有得到俄罗斯宫殿的版权许可，以专业的日本摄影师的视角为蓝本的图片，不是依原样复原其中世纪的面貌，而是经过编辑、整理覆盖于墙纸之上，呈双层的态势，鬼魅的意象。特有的神采飞扬，华光般闪耀在本案幽幽的调色板。

昔日经典般的一个空间，绝对为青年人钟爱。火爆的饮料销售、热闹的人群是极好的明证。

This club lounge, Rootz, was planned in Kuala Lumpur Malaysia, the roof parking space of the existing shopping center named Lot 10 which greets the 20th anniversary. I felt fortunate as I was able to participate and design for the master plan which included a club, theater, restaurant and the place of the courtyard.

I resembled the rooftop of the Lot 10 shopping centre and making a club lounge as I thought vaguely of different region of people mingled around and dance together in the grand hall of the palace, because there was the certain figure from the old days whether people from the east or west could dance together.

The plan was adopted soon, so I negotiated patiently with the different palaces in various countries and took permission for the copyright from the Russian palace and finally the photograph was taken by a photographer from Japan together with me. We edited it and printed it on form to fit a new skeleton as wallpaper. Thus, we printed it on an organza and hung it in front of the wallpaper to make double layers and represented a ghost phenomenon. I edited it instead of restoring the palace of the middle ages exactly and not to become too artificial so that the dimensions will only encountered accurately. Then, I had a thought across my mind to create a bright illuminated club to against the typical dark black club.

Even if it is a bright club lounge, this club is absolutely accepted by the youth. I feel so relief after knowing that the record explosive sales, drinks sold out, and 200 people whom not be able to enter and waited patiently for their turn in the courtyard.

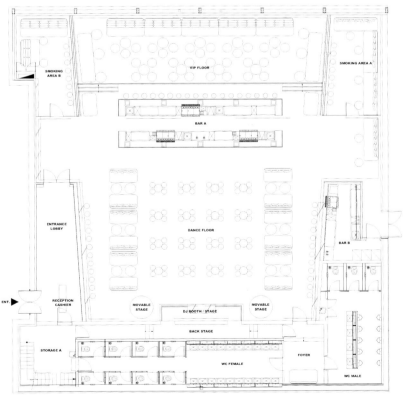

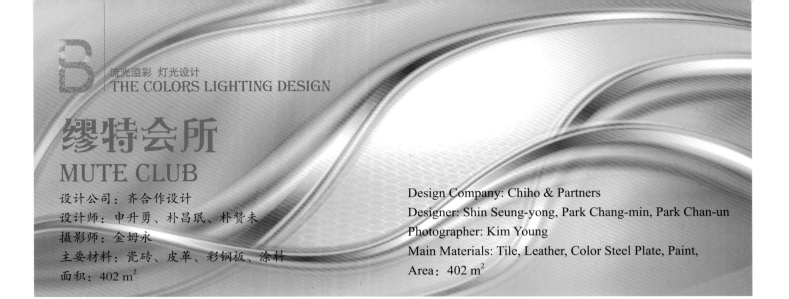

缪特会所
MUTE CLUB

设计公司：齐合作设计
设计师：申升勇、朴昌珉、朴赞未
摄影师：金姆永
主要材料：瓷砖、皮革、彩钢板、涂料
面积：402 m²

Design Company: Chiho & Partners
Designer: Shin Seung-yong, Park Chang-min, Park Chan-un
Photographer: Kim Young
Main Materials: Tile, Leather, Color Steel Plate, Paint,
Area: 402 m²

梨泰院集东西方文化之大成，但同时又是韩国最富异域情调的区域，此言毫不为过。

缪特会所位于"汉密尔顿大酒店"一、二楼空间。设计融万千风格、文化为一体，但却又尽展梨泰院风情。

一、二楼空间有爱尔兰风格的"快乐"酒吧，有欧式传统的"迷人"酒吧，也有令人回想起美国20世纪二三十年代的"缪斯会所"。

Being the concentration of Eastern and Western culture, it is not too much to say Itaewon is the most exotic area in South Korea.

The District, located on the first and second floor of the Hamilton Hotel, was designed as a complex cultural space holding the uniqueness of Itaewon, where a diversity of styles and culture coexist.

The project occupying the first and second floor includes Irish pub "Prost", traditional European lounge "Glam", and club "Mute" evoking on the 1920—1930 of the United States of America.

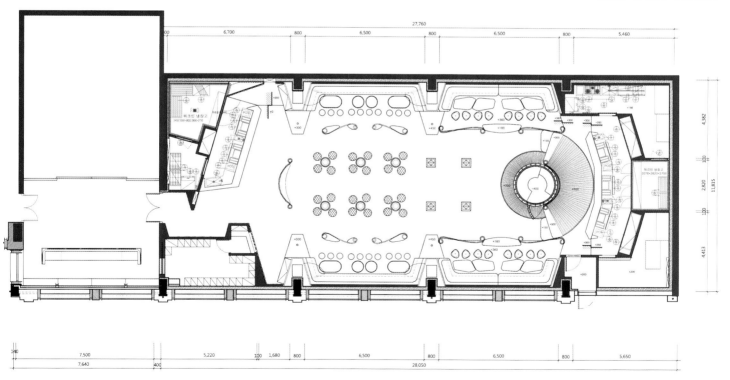

流光溢彩 灯光设计
THE COLORS LIGHTING DESIGN

烟尘会所
THE SMOKEHOUSE ROOM –NIGHTCLUB

设计公司：巴士拉伊德设计工作室　　　　　　　　　　Design Company: Busride Design Studio

　　巴士拉伊德设计工作室曾于印度新德里设计过一个饭店，名为"烟尘大酒店"。本案从量体上而言，可以说是其附属的一部分。建筑上的关联，自然而然地使其得名为"烟尘会所"。

　　1 115平方米的酒店空间，有机延展，俯瞰着库特博文化遗产风景区。该区域景观可谓是印度最为壮观的区域。借助于设计妙手，"烟尘大酒店"以其后现代的透视棱镜，架构着遥远的13世纪的历史。

　　3个不同的区域，无缝对接，任由外部的风景以流性、有机的状态融入空间内部。

　　统一于新德里的新兴会所中，"烟尘大酒店"给予人的是完全沉浸式的"视听"环境。成功地给人一种迷离、兴奋、神奇的体验。"蘑菇"式的天花，正欲绽放于内外之间。伴着灯效，脉冲式的视觉，抽象的心理感受似乎有了具象的表达。渐进式的EMD声响系统，响彻于内外，空间因此有了一种呼吸的冲动。印度本土的会所文化于是变得更加国际化。

　　"烟尘会所"，是能量的迸发，是青春的气场。投影地图墙创造着"视听"体验，孕育着氤氲的会所气氛。定制的罩灯，在宽阔的空间里，投下戏曲的张力。

　　会所最上部的VIP室，俯瞰着舞池，是空间最独特的区域。升华着，加剧着烟尘俨俨的气息。

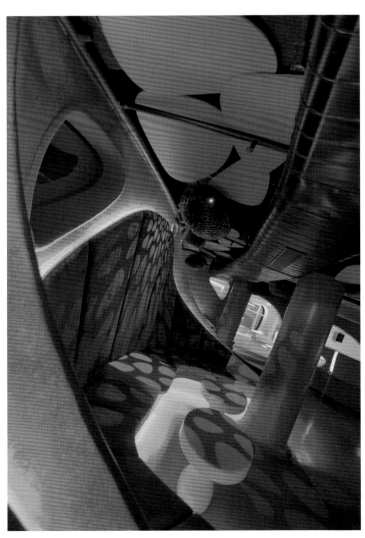

Busride Design Studio have designed a restaurant called The Smokehouse Room as well as an attached nightclub called SHRoom in New Delhi, India.

One of the most bizarre briefs we've worked with, Smokehouse Room flows organically out over 1,115 square meters of curves, overlooking the grandest view in India, the Qutub Heritage precinct. The Smokehouse Room frames 13th Century history in postmodern lenses.

The Smokehouse Room has 3 distinct, yet seamlessly connected offerings. We've tried to create a fluid, organically growing, psychedelic landscape that melts into various parts.

The Smokehouse Room is attached to the newest entrant to the Delhi Club circuit, SHRoom. With SHRoom we created a completely immersive Audio-Visual Environment, in what is intended to be the closest replica of the psychedelic mind-bending experience. The club tries to visualize an experience inside and under an exploding canopy of mushrooms, with synced effects lights and pulsing visuals. SHRoom adopts the progressive sound of EDM to create a space that breathes and lives, pushing the Club culture in the capital city into a more international space.

The Nightlcub, SHRoom, attached to the Smokehouse Room forms the high energy extension of the concept, creating a young, buzzy atmosphere furthering the spirit of experimentation. The Projection mapped walls at SHRoom allow for a seamlessly integrated audio-visual experience, and an immersive club atmosphere. Custom shadow lamps at SHRoom, allowed us to create drama across the large expanses of the club

The most exclusive area, the VIP lounge at the head of the club, overlooks the entire dance area. All these can be called the immersive club experience at SHRoom.

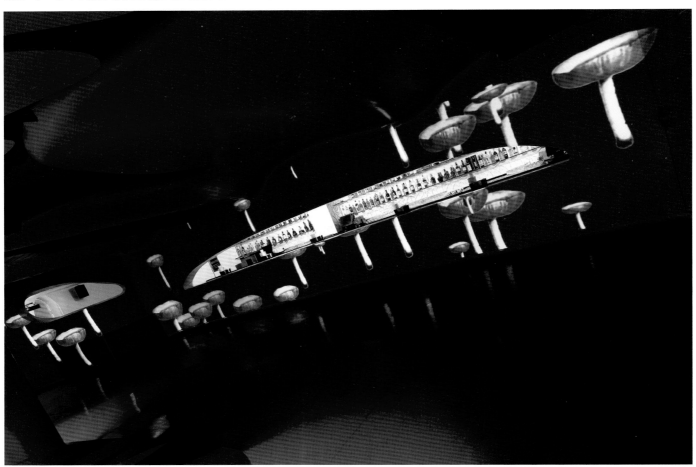

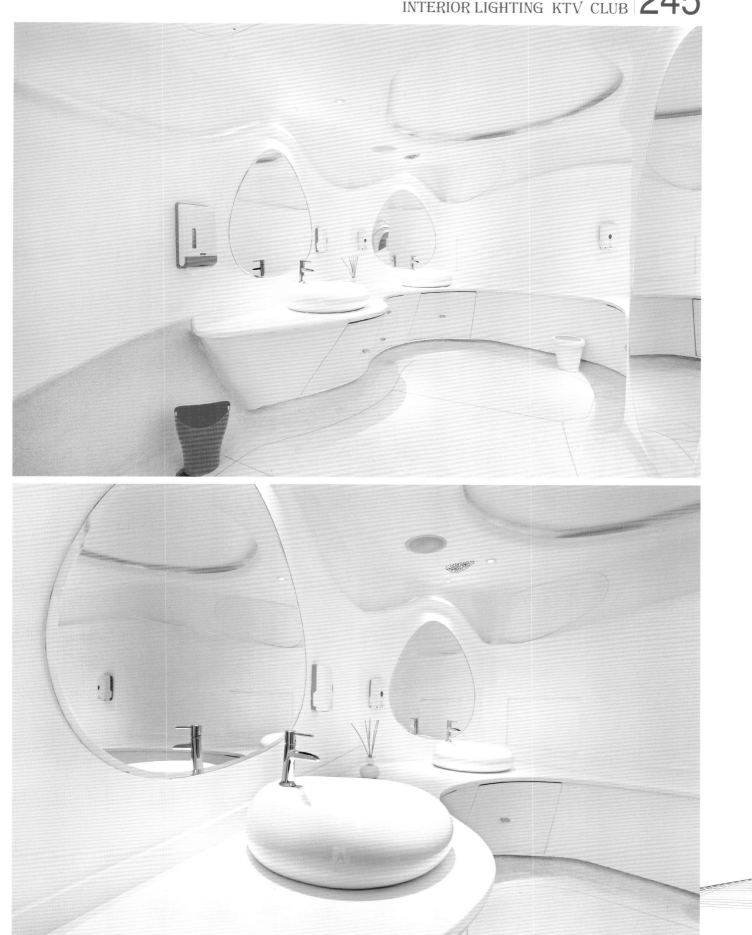

B 流光溢彩 灯光设计
THE COLORS LIGHTING DESIGN

北京麦乐迪KTV 月坛店
MELODY KTV (YUETAN, BEIJING)

设计公司：睿智汇设计公司	Design Company: Wisdom Space Design
照明设计：睿智汇设计公司	Lighting Design: Wisdom Space Design
设计师：王俊钦	Designer: Wang Junqin
参与设计：彭晴、樊旭、张庭	Participate: Peng Qing, Fan Xu, Zhang Ting
摄影师：孙翔宇	Photographer: Sun Xiangyu
主要材料：浅啡网石材、黑金花石材、黑镜、粉镜、玫瑰金镜面不锈钢、人造罗马刚石、LED灯、钢化清玻璃、金属马赛克	Main Materials: Marble, Black Mirror, Pink Mirror, Rose Gold Stainless Steel, Corundum, LED, Tempered Clear Glass, Metal Mosaic Tile
面积：2 700 m²	Area: 2,700 m²

灵感由来：灿烂的烟花。
主题思想：烟花的多变及绚丽璀璨。
风格定位：中高端商务人士娱乐及商务会议/用餐结合之场所。

布局设计：公共区由大厅/服务区/精品超市及会所构成，继而为主次走道及大中小包厢和主题包厢。

这个案为第二次的设计整改，因而以烟花绽放为设计主题。也意指个案的再次绽放，全新包装全新出击，给顾客新的感受及娱乐环境。

公共区由大厅、服务区、精品超市及会所构成，继而为主次走道及大、中、小包厢和主题包厢。定位于中高端商务人士娱乐及商务会议、用餐结合之场所。以"绽放"为设计主题，将绽放的元素贯穿个案的整体，烟花总是绚丽璀璨并多彩的，以LED灯的色彩变化取代烟花绽放时的灿烂及多彩，各种色系的不锈钢材质镜面反射呈现烟花绽放一瞬间的多样变化图案，而烟花总是在夜晚的释放才显得更具魅力。因此，设计主题的另一色系即为黑色，此案运用了黑金花石材表现夜晚及星光，更显出烟花绽放的主题。

Inspiration: fireworks.
Theme: The change and brilliance of fireworks.
Position: high-end, for entertainment, business and dining.
Layout: public area of hall, service, and boutique, aisle, box and theme box.
The design for this project is actually a reformed design, where to take fireworks bursting into bloom, a symbol of this project bound to be flourishing and prospering with a completely-new packing that allows for a new experience and a physical entertainment setting.
The public consist of hall, service area, boutique supermarket and chamber, while aisles links boxes of different sizes and theme boxes. The space positioned high-end for business circle to have meetings or dine takes flourish and prosperity as its themes where to carry through relevant elements of fireworks, whose brilliance is highlighted and set off with adjustable and changeable LED. Meanwhile, mirror stainless steel reflects the diversity of pattern at the very moment. Due to the time of fireworks, the hue of black is given to full play, particularly the black coated in black color make the spatial theme more prominent.

流光溢彩 灯光设计
THE COLORS LIGHTING DESIGN

麦乐迪南京新街口店
MELODY KTV (XINGJIEKOU, NANJING)

设计公司：睿智汇设计	Design Company: Wisdom Space Design
照明设计：睿智汇设计	Lighting Design: Wisdom Space Design
摄影师：孙翔宇	Photographer: Sun Xiangyu
主要材料：水钻、玫瑰金镜面不锈钢、茶镜、人造石材、镜面不锈钢激光冲孔、马赛克镜面砖	Main Materials: Diamond, Stainless Steel, Tawny Mirror, Stone, Laser-Punched Stainless Steel, Mosaic Tile
面　积：4 000 m²	Area: 4,000 m²

南京，历来就有"六朝古都"的美誉，透露着儒雅之气，豪杰之风，斯文秀美，而当文化与娱乐交织，历史与激情结伴，又会激发怎样的碰撞？

它是顶级娱乐空间的代表，是激情与魅力的代名词，是这个古老城市商业街中诞生的璀璨明星。

本案是中国著名娱乐连锁品牌麦乐迪的KTV直营店，坐落于南京市新街口商业区，面积为4 000平方米，市场消费人群定位于都市男女，是睿智汇设计公司自麦乐迪KTV中服店设计后的又一力作。总设计师王俊钦在设计过程中大胆地运用对比手法，以精致典雅却又不失现代气势的独特气息另辟蹊径。

挑高中庭是本案的设计重点，设计师将盛开的"莲花"置于半空中，作为视觉中心点。"莲花"的设计取用了莲花本身的美好含义，运用了后现代设计风格中的隐喻手法，代表着超脱幻象新世界的诞生。"莲花"采用了玫瑰金不锈钢材质打造，将莲花的柔美与金属材质进行对比与碰撞。莲花取水于源，将"花"与"水"相互衬托，相映成趣。"水"作为主体莲花的背景，以流线型的设计语汇呈现于吊顶及主墙。顶面造型采用镜面不锈钢材质，配合晶莹透彻的水砖，精美绝伦的闪耀呈现，在灯光下闪闪发亮，有着目眩神迷的造型和闪耀潮流的样貌。设计师将水的造型处理成旋转的、波浪起伏的条带，在虚与实、轻与重、固定与流动、开放与封闭、光泽与透明之间，营造了一个抽象、动态、迷幻的空间感受，给人一种视觉的享受和联想，振奋人心。

麦乐迪KTV的空间设计是着重于艺术化与商业化的完美结合，不断捕捉消费群体对生活方式的追求，探索如何提供全球最独一无二的娱乐感受，不断打破消费者和业主设计期望，创造出这令人惊艳的璀璨之作，同时成就了这个古老城市的熠熠生辉，带来全新的娱乐气息。

Nanjing, a city honored as a capital of 6 dynasties, is gentle, bold and forthright in terms of temperament, but equally beautiful or scenic with respect to its landscape. Then what will become of such a physical setting, if collision takes place of culture and entertainment, and history and enthusiasm?

And Melody, a representative of the top level recreation as well as passion and attraction, is bound to be a star already rising out of the prosperity of the ancient business area.

The project, another shop of China's renowned brand, covers an area of 4,000 square meters, a space positioned urban people as its customers that make a sharp but intended contrast between graceful elegance and fashion beneath the skillful hands directed by Chief Designer, Mr. Wang Junqing.

The atrium of raised height is the spatial focus, where lotus in blossom is suspended in the air, a metaphor approach of post modern design that overspreads the good connotation of lotus and embodies the birth of an unconventional new world. The stainless steel of the lotus sets against the morbidezza of the plant. Water now serves as the lotus backdrop. Over the walls it's stream-lined. Suspended to accomplish fun by water and flower, the ceiling is of mirror and stainless steel, shining against the crystal brick. The water weave is sometimes open, but then closed, or sometimes static but suddenly becomes dynamic, making visual effect abstract, flowing and fantastic. All allows for enjoyment, imagination and freshness.

A project the Melody KTV is where to perfectly combine the commercialization and art, constantly capturing consumers by providing them the most unique entertainment experience only available here. The stunning surprise repeatedly gone beyond expectations creates a new entertainment atmosphere in the ancient city of Nanjing in glittering freely at its wishes.

流光溢彩 灯光设计
THE COLORS LIGHTING DESIGN

广东中山魅力皇爵KTV会所
CHARM BARON KTV CLUB (ZHONGSHAN, GUANGDONG)

设计公司：深圳品彦室内（娱乐）设计有限公司
设计师：杨彦

Design Company: Shenzhen PinYan Interior Design Co., Ltd.
Designer: Yang Yan

夜幕降临，灯火初上之际，一颗璀璨夺目之星华丽登场，它是所有最时尚的、最前卫和商务人士最爱的聚集之地，夜总会的出现让他们眼前一亮，紧随着是无比的雀跃、快乐和兴奋。这就是他们渴望的无忧空间的快乐的源泉。

魅力皇爵的空间不是简单的室内装潢设计，更融入了前沿的创新、美观、实用和文化元素理念，奇妙的动态导向设计，金碧辉煌的大堂，低调奢华的空间感，时尚新颖的造型，加上魔幻的光线，直接让人感受皇爵的高贵和奢华。让人彻底忘记压抑，以最快的速度进入快乐之旅。这正是皇爵的魅力所在。

本案的设计理念大胆创新，设计师以节能环保为本，结合空间美学和文化元素，力求打造一个美观、时尚、有内涵的夜总会！在灯光设计方面，本案打破常规夜总会采用直线光源照明的模式，而采用暖色调漫反射照射模式，合理利用天花、墙面和地面的材料本身反射光源。从美学角度上让人感觉仿佛进入巴黎的卢浮宫，体现本案的高贵和华丽，最重要的是漫反射光源照射的灯光模式真正实现了节能最大化。

房门的设计理念是引用基督教中一座拥有500多年历史的教堂大门的元素延伸而来，西方教堂的文化与中式夜总会，看似两个永远不会产生交集的文化，我们却大胆将它们巧妙地结合在一起。配合灯光和装饰材料，它们被赋予了新的生命，成为了承载着该夜总会场所的国际性、美观性和文化的基石。

总统房的设计，在低调奢华的基础上也注入了诸多文化元素。天花上的手工壁画，灵感来源于设计师对古文化的研究和探索，敦煌莫高窟是我国珍贵的历史文化遗产，为我们留下了博大精深、内容极为丰富的文化艺术宝库。在21世纪的今天，在打造最时尚前卫的休闲夜总会的时候，我们更是可以结合古文化和古艺术，让古今文化交融，将中国传统文化延续、传承和发扬！

As night sets in, a dazzling star debuts gorgeous, making itself to be the most fashionable, avant-garde for business people. Refreshed and happy, they realize what they want to get in an entertainment space here.

Not a mere interior design, Charm Baron has now fused elements of innovation, aesthetics, function and culture. The guidance design in a dynamic state, the resplendent lobby, the low-key luxury, the novel shape, and the magical light contribute to a direct exposure to nobility and extravagance. When completely forgetting the ever suppressed, people can access to a happy journey at a fastest speed. This is what the charm of the space means.

With a concept of bold innovation, the space is oriented environment protection when combined with spatial aesthetics and cultural elements to build a beautiful, fashion, and meaning space. Lighting break away the normal procedure by using linear lighting, warm color to reflect exposure, and surfaces of wall and ceiling to mirror light. Here makes another Le Louvre in terms of aesthetics, embodying the noble and magnificent. And what's the most important is that such a diffuse reflection really maximizes energy saving.

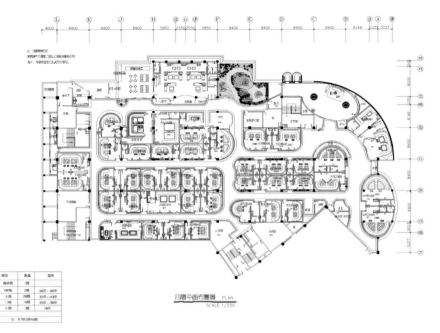

The door design is inspired by a gate of a 500-year-old history, which bridges the quite two different elements of Christian church and Chinese nightclub, elements that have now been given to a new life with lighting and decoration material. A cornerstone it is carrying international taste, beauty and culture. Based on a low-key luxury, the president suite is implanted with many cultural elements. The ceiling manual murals, inspired designer's personal study and exploration of ancient culture, particularly Dunhuang Mogao Grottoes, a precious historical and cultural heritage that has left broad and profound contents. In a setting of 21st Century, a most avant-garde club combined with the ancient culture and art is an illustrative example to inherit and carry forward traditional Chinese culture, inherit and develop!

室内灯光　KTV　会所
INTERIOR LIGHTING　KTV　CLUB　263

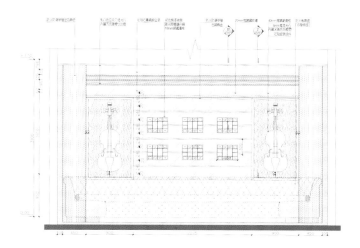

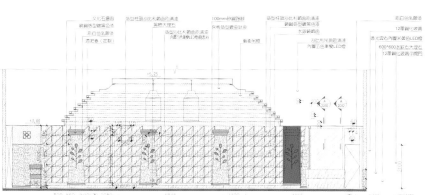

流光溢彩 灯光设计
THE COLORS LIGHTING DESIGN

咏歌汇 2 期
PHASE 2 OF YONGGE HUI

设计公司：罗一博装饰设计有限公司
设计师：罗国春

Design Company: Luo Yibo Decoration Co., Ltd.
Designer: Luo Guochun

根据地区特点，设计师确立了与众不同的手法和材料。本方案只针对高端客户，一改以前昏暗、低沉的娱乐场所设计风格，采用了明亮、典雅、奢华的设计定位，广阔的流通路线，双向的上下楼通道，材料多以现代、明快的材料为主。壁画、水晶运用比较多，由于地域原因，定位为高端消费客户，设计新颖、奢华，会所提升了一个档次，经营后形成了良性循环。

Any interior design, if good, would adjust to the regional characteristics with distinctive techniques and materials. So does project positioned high-end in breaking away from the stereotype of entertainment industry, no longer dark or gray, but bright, elegant, and luxurious. Circulation routes are broad, with two-way stairs bridge up and down. Materials focus on the equally bright, like painting and crystal used in amounts. The physical location contributes to its high-position, and the innovative and luxurious design is bound to raise the price level and operation to an incomparable grade, forming a virtuous circle after being put into use.

流光溢彩 灯光设计
THE COLORS LIGHTING DESIGN

康业国际会所
KANGYE INTERNATIONAL CLUB

设计公司：罗一博装饰设计有限公司　　　Design Company: LuoYibo Decoration & Design Co, Ltd.
设计：罗国春　　　　　　　　　　　　　Designer: Luo Guochun

　　与同类竞争性物业相比，作品独有的设计策划、市场定位，是酒店投资方设立的配套服务，同时酒店经营方面也实现了可观的效益。与同类竞争性物业相比，作品在环境风格上的设计创新点，设计师采取解构主义设计手法，加上变色LED灯结合创造出一种耳目一新的空间感受。作品在空间布局上的设计创新点，合理利用建筑空间，经营面积最大化的同时充满色彩感，与同类竞争性物业相比，作品在设计选材上的设计创新点，每款包房墙面的设计都是独一无二的。包房巧妙地运用了玫瑰金、水晶胶、LED灯无处不体现着现代专业气息，冲击到访者的每一个细胞。

Compared with the competitors, only with unique design planning, market positioning, and supporting services, can a project be everlasting while hotel operations is able to achieve considerable benefits So does this project to bring out refreshing feelings out of the space with deconstruction approach. Design technique innovation leads to a reasonable spatial use while maximizing the sense of color for the operating area. Besides all walls in each private, the employment of rose gold, polyester resin, and LED, overspread the modern air everywhere, making a strong strike on visitors.

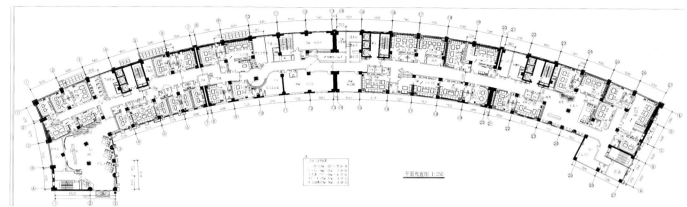

流光溢彩 灯光设计
THE COLORS LIGHTING DESIGN

天命夜总会
KISMET NIGHTCLUB

设计公司：黑羊设计
设计师：蒂姆、马克、本
面积：3 450 m²

Design Company: Blacksheep
Designer: Tim Mutton, Mark Leib, Ben Webb
Area: 3,450 m²

印度有个城市，名叫海得拉巴，是个动感之城。它充满活力、能量，正以积极的姿态向前发展。海得拉巴有个酒店，名为公园大酒店。酒店临湖而居，尽享湖景水色。本案夜总会，有1 068平方米空间，位于公园大酒店泳池平台下。其空间设计，如其所在城市，充满动感，富有活力。

二层空间内设VIP酒吧、主酒吧、俱乐部和贵宾室。整体空间风格与公园大酒店一脉相承，传承过去，继往开来，尽展现代城市美学。

设计整体手法源于海得拉巴地区传统美学、工艺，同时融之于现代家具、铺陈。所有用材设备，取之于当地，制作于当地。原汁原味的当地风情，不仅于表，更蕴于里。

用材色泽不一，或金属质感，或宝石光泽，或彩虹般闪耀。丰富的色度层次，另有"珠宝"的映像。夜店的丰富多彩真正实现了多角度的展现。通往内里的走道，墙体饰以小面的不锈钢板材，内嵌的照明，连接着DJ声控系统。

主空间层次更为丰富。后区自区抬高，设计如剧场的观众席。舞池配备音乐同步照明。舞池四角，各以席座。各席座上空，以定制吊灯作为悬挂。吊灯由传统的镀金绳索、珠子拧结而成。所有用材全部由当地市场采购。

巨大的中心酒吧具有雕塑般的质感，气势流畅，连接着VIP区域及主空间。闪闪发光的可丽耐人造大理石内嵌着灯光照明。两个区域也因此成了空间的焦点。

半透明的亚克力墙界定着VIP与其他空间的界限。两层的树脂层中空设计在滤光的同时，又成为隔音的屏障。从室内的天花至泳池的地板之间，无不体现着一种"虚空"的感觉。那其实是空间的特色：量体的虚空，正等待着俨俨的气氛来狂欢。

通往主俱乐部空间的地方，一片翡翠般的清爽区域。其中一堵富有特色的图案墙，以华丽丽的姿态，用现代的用材诠释着当地的传统。秋千式的摇椅，为空间增添了儿时的纯真与快乐。

世界级的夜总会，可以同印度业内任意一家单位相媲美。本案是海得拉巴的夜空一个令人向往的头号狂欢之地。

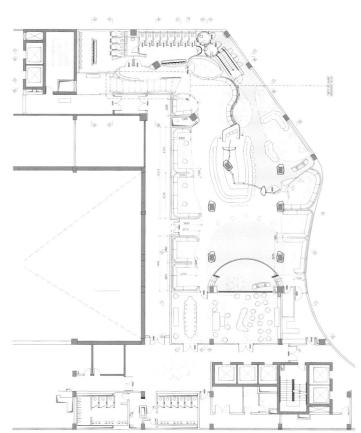

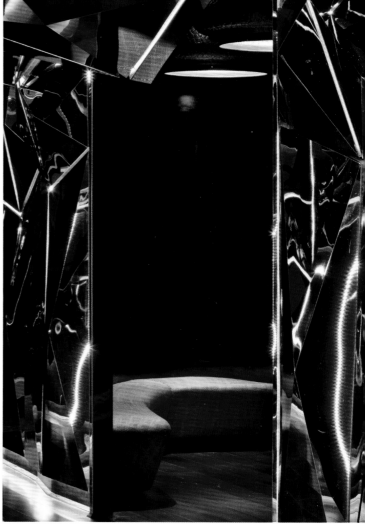

Reflecting the energy, drive and optimism of Hyderabad, a dynamic city, Kismet is a spectacular 1,068 square meters nightclub set beneath the pool deck of The Park Hyderbad—a new-build hotel on the city's exclusive lakeside.

The second-floor venue comprises a VIP bar, main bar, club space and a private function room. In keeping with the rest of the hotel's bold and highly contemporary architecture, Blacksheep's design transports visitors into a fantastical and escapist environment where respectful nods to the past are seamlessly woven into an ultra-modern aesthetic. The overall design approach, was to refer the wonderful artistic and craft traditions of the Hyderabad region, but to fuse these elements with highly contemporary treatments, furniture and features, reflecting the city's hi-tech present-day reality. The materials and commissions displayed within Kismet have been sourced or fabricated locally.

References to jewellery are particularly evident in an interior treatment layered with metallics, gem shades and iridescent materials. Visitors enter the club, for example, via a spectacular tunnel, the walls of which are clad in faceted stainless steel panels, with inset lighting linked to the DJ's sound system.

The main club area offers plenty of circulation space on arrival, with

a raised "people-watching" seating area at the rear. The dancefloor, equipped with music-synchronised lighting, has four large booths, one at each corner, each lit by an illuminated bespoke chandelier made of traditional gilded ropes and beads-sourced from a local market.

A huge-scale central bar runs through both the VIP area and the main club. Smooth, sculptural and made of glistening Corian with inbuilt lighting, it provides a dramatic focal point for both areas.

A semi-transparent acrylic wall separates the club from the VIP space. Made up of two layers of textured resin with a gap in between, it provides a soundproofing barrier while allowing a certain amount of light to filter through. The VIP area also boasts what is arguably the club's most dramatic feature: a void leading up from its ceiling to the floor of the swimming pool above, creating a truly unique effect.

Leading off the main club space, the chill-out area is accented with tinges of green to provide a sense of calm, while a feature wall reinterprets traditional ornate local patterning in contemporary materials. A swing seat adds a note of childlike fun.

The overall result is a world-class nightclub that not only established itself instantly as Hyderabad's number one nocturnal destination but that rivals anything India had to offer.

胜悦国际头皮养护馆台北市南京路店
SHENGYUE HAIR CARE (NANJING EAST ROAD, TAIBEI)

设计公司：城市室内装修设计有限公司　　Design Company: City Interior Decoration Design Co., Ltd.
设计师：陈连武　　　　　　　　　　　　Designer: Chen Lianwu

　　胜悦国际头皮养护馆位于台北市南京东路精华地段，占地面积660.756平方米，是一座四层楼建筑物，包括B1未来涟漪层，1楼中介涟漪层，2楼当代涟漪层，3楼过去涟漪层。

　　空间以带动时尚的连锁效应，犹如时空涟漪，色即是空的概念贯穿整个设计，搭配热情红色的运用，让发型空间就和头发一样多变。

　　走进一楼中介涟漪层，就像走进时空隧道，随意穿梭于前、后台，可以感受表演者与观赏者身份瞬间交替；空间错位、角色对调，让人充满期待。吧台区墙面以红色涟漪墙面带动欢愉的热闹气氛。等待观众席则以白色墙面为背景，投射出红色灯光的时尚氛围。楼梯间，天花板以白色放射状涟漪，延伸贯穿整个楼梯。

　　B1空间以科技感呈现，属于未来涟漪层，墙面上硕大的面板，利用数字密码中0与1的无限思考，黑与白的极简利落，神秘而迷人，完全展现对于未知世界的探索欲望与想象力。

　　二楼为当代涟漪层，以生命必需物质——金、石、水、火、土、阳光、空气、水为形。水区中潺潺的水幕与光透的墙面，水波粼粼的地板，似大小水滴的吊灯垂落散置，给人晶莹剔透的清新感受。温室植生墙与错落的林木造景，仿佛进入奇幻森林一般，茵茵绿叶环绕在身旁，让您有置身原始森林的错觉。吐纳之间恣意享受天然清新气息；搭配轻柔谐和的大地色系，现代感透视镜面设计，营造开阔视野；暖色系灯光笼罩，如阳光般闪耀，让您化身原野中的精灵，远离都市喧嚣。

　　三楼为过去涟漪层，以过往情境组构出老街、老房间、老故事与旧梦四组空间。

室内灯光　其他
INTERIOR LIGHTING OTHERS

The project takes up a golden positon of Nan Jing East Rd, Taibei. The area of 660.756 square meters covers a 4-floor building, B1 Future Ripple, the 1St Medium Ripple, the 2nd Today's Ripple, and the 3rd Yesterday's Ripple.

A space it is that takes fashion to stir up a chain effect, like space-time ripple, where the concept of "Infinite" is carried out throughout with red to overspread zeal and passion, so that a saloon can be changeable as much as hair.

The 1st floor, Medium Ripple, makes a time tunnel. Around desks front and back, the instant switch from performer to viewer can be felt. The spatial malposition and the character change lead to expectation and recognition. The wall around the bar counter is patterned with rip to compliment the joyous atmosphere, the wall of the auditorium coated in white with red lighting projecting fashion. The ceiling of the stairwell is embellished with ripple that goes around the whole staircase.

The B1 space, the Future Ripple, is endowed with technological sense, where Qr Code is inlaid on the wall. The thought on the digital password of zero and one is optimized. The black and white is charming and mysterious, fully embodying the exploring desire and imagination of the unknown world.

The 2nd floor, Today's Ripple, takes life as its substance carrier by utilizing gold, stone, water, fire, earth, sunlight and water to develop forms and shapes. The water screen, the illuminated wall, the wavy floor and the water-drop chandeliers allow for refreshing feelings: crystal and clear. The greenhouse wall and the well-arranged trees seem to have made a forest fantasy. With green leaves everywhere, you feel nothing but an illustration that you have been in a primitive forest, breathing the fresh air. The dominant hue is earthy. The mirror design is modern and transparent. All together accomplishes a wide range of view. Lighting fixture casts warm, sunlight like to make you an elf in wild with urban bustling and hustling kept far away.

As for the 3rd, it's named as Yesterday's Ripple, where situations past are made, with old streets, old rooms, old stories and old dream taken as the theme to set up four kinds of sections.

B 流光溢彩 灯光设计
THE COLORS LIGHTING DESIGN

Kippo 美发沙龙
KIPPO HAIR & COLOR BAR

项目位置：澳大利亚新南威尔士
设计公司：Vie 设计机构
摄影师：莱伯里克
面积：86.4 m²

Location: Sydney NSW, Australia
Design Company: Vie Studio
Photographer: Laeberlinc
Area: 86.4 m²

Kippo Hair & Color Bar 能够提供独一无二的豪华体验，满足人们对零售的需要，为现代、专业的美发沙龙树立一个典范。为了提高酒吧在这个行业中的地位，Kippo 致力于给自己时尚前卫的顾客打造一个时尚、个性张扬的空间。

由邻近租客的限制，Kippo Hair & Color Bar 希望用 86.4 平方米的有限空间创造一种空间无限的感受。Kippo 力求通过对材料和灯光的运用来打破场地现状达到一个新的高度。

由于现有铺面的原因，场地墙壁和结构必须保持不变。装修时的负面影响尽可能减少，从而确保预算的控制。洗发室现有的墙保持原状，以此与设计所用元素保持一致。

Kippo Hair & Color Bar is the answer to a modern and slick hair salon that offers a unique and luxurious experience to suit not only the highly popular retail precinct but also its demographic. Seeking to raise the bar in this industry, Kippo aims to push new boundaries in design by creating a stylish and indulgent space for their fashion savvy patrons.

Constrained by neighbouring tenants, the confined space of only 86.4 square meters subjected to a design that created a perception of infinite space. The design of Kippo seeks to reach new heights beyond the site's existing physical boundaries with the play of materials and lighting.

Due to existing shopfront, walls and structures within the site were required to be remained intact, minimal construction work was a significant factor which positively impacted the project budget. The existing walls of the shampoo rooms have been retained in keeping with the raw elements of the design.

室内灯光　其他
INTERIOR LIGHTING OTHERS

B 流光溢彩 灯光设计
THE COLORS LIGHTING DESIGN

意念空间
THE ROOM

项目位置：意大利米兰	Location: Milan, Italy
客户：GR8	Client: GR8
设计公司：莱工作室	Design Company: LAI Studio
设计师：毛里齐奥	Designer: Maurizio Lai
主要材料：金属网、玻璃、木板、金属板、高密度竹板、中密度彩色玻璃板、定制墙饰、LED照明	Main Materials: Metal Mesh, Coupled Glass, Wooden Panels, Metal Plates, Hi-Density Bamboo Boards, Medium-Density Colored Fibre-Board, Custom Printed Wall Coverings
面积：350 m²	Area: 350 m²

"意念空间"位于米兰，是一个全新的行政级别的俱乐部，开放时间为晚8点至凌晨2点。整个空间弥漫着"纽约客"的气氛。艺术般的文化气氛中，虽然以一种街头食物般的理念诠释，却也写尽了优雅与别致。一个空间，多样面。国际化的品位、艺术与天分尽情展现着一个城市的美。350平方米的空间附设有一个小小的室外卡座。两层的空间因为其颇有震撼力的楼梯与背光的酒吧显得突出而醒目。

The Room is a new and exclusive club located in Milan's Porta Romana, which is open from 20:00 to 2:00. New Yorkers atmosphere, a wide range of artistic and cultural evenings, accompanied by a reinterpretation of the concept of Street Food in a more elegant and refined key, making the Room a multifaceted one place, where to discover the beauty of a city with an international flair made of taste, art and talent. With an area of 350 square meters and a small outdoor seating area, The Room is structured on two levels, in which stands out for its architecture the imposing stairs and the backlit bar, as well.

B 流光溢彩 灯光设计
THE COLORS LIGHTING DESIGN

水族馆酒吧
AQUARIUM BAR

设计公司：新起步工作室	Design Company: Next Level Studio
设计师：米其	Designer: Michal Kutalek
参与设计：维克多	Participate: Viktor Johanis
摄影师：马丁	Photographer: Martin Kocich
主要材料：纤维、硬纸板、环氧地坪	Main Materials: Sideglowing Fibre, Chipboard Panel, Epoxide Floor
面积：70 m²	Area: 70 m²

"水族馆酒吧"位于联排别墅的后部，为后来新增加的部分。该联排别墅的历史可以追溯到20世纪90年代。新添加的空间把整个地基都占完了。旧时的天窗就成了采光的唯一通道。

淡淡色调的空间以纤细的纤维作为肌理。3毫米的纤维穿过12个玻璃框架，决定了整个空间的形态。所有的纤维加起来，总长度超过了1千米。两个光源运用于空间，整体色度渐进中溢满空间。光感变化的速度可以人为控制或加以固定。参数式的设计方法运用于结构之中。电脑中的软件根据需要与不同形状变量相互融合。基本的参数变化决定着玻璃框架的直径与长度。框架内的纤维数目等等决定着框架的半径。参数设计方法的使用不仅决定着形状，还决定着玻璃框架的生产草图以及纤维的长度。

6米宽的水族馆是本案名称的来源，如今已然融合进了前部空间。轻盈的结构却创造了一个主要的主导空间，弥补了街景缺失的遗憾。精致的纤维网集美学、功能于一体，阻挡着来自外界的声音。分量配置的LED灯带置于玻璃的框架上，赋予空间更多的色彩变换。

The area of cafe-bar is located in the back additional building of town house, which dates from the 90s of the 20th century. Additional building fills the whole

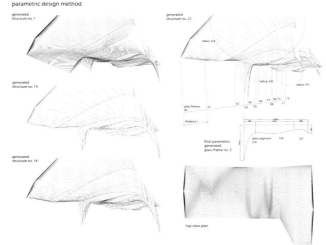

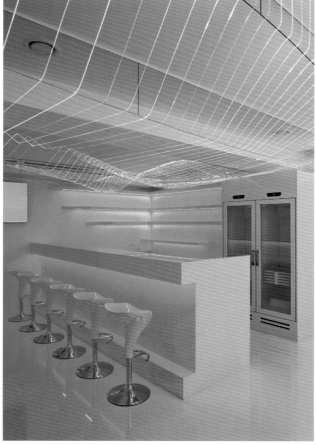

...round and therefore the only source of daylight is the original skylight.

The space is light colored structure of light-conducting fibers. The fibers with a diameter of 3 millimeters through 12 glass frames that determine the final morphology. Total has been used over 1000 m fiber and two light sources that create gradual spillover colors through fibers. You can regulate the speed of color change or suspend it in a certain tone. The structure was designed using a parametric design method, where based on our creating of software tools, were generated different shape variations after entering certain parameters. These were examined in visualization and then changed again. The basic parameters that can be changed are the dimensions and distance of glass frames, their radius controlled by ripple control circles, the number of fibers passing through the frame, their densification, etc.

The method of parametric design served not only for finding the optimal shape, but also to easily create production drawings glass frames and accurate determination of the required length fibers.

The new name "Aquarium" is derived from 6 m wide aquarium, which is integrated into the front panel. Together with light structure create a major dominant of space and compensate for missing view of the street. Fine fiber network also serves as a breaker sounds. RGB LED strips deposited on glass frames are used for any illumination and coloring space.

B 流光溢彩 灯光设计
THE COLORS LIGHTING DESIGN

咖啡酒吧
CAFE BAR

设计公司：新起步工作室
撰文：米甲
合作：维克多
面积：67 m²

Design Company: Next Level Studio
Text: Michal Kutalek
Cooperation: Viktor Johanis
Area: 67 m²

本案业主在邻近社区寻得了此处有点类似于乡村别墅般的空间。设计师希望在不长的工期内，以建筑的理性于酒吧内呈现一个咖啡空间。

酒吧位于空间前部，视野独好。吧台钢构覆盖着纸板，外以涂漆。有些功能设计隐于吧台之内，并与侧墙相接。前面墙体设有壁龛，专门用于美酒与器皿的展示。黄色的霓虹照亮着漆黑的内里。高光的反射夸大着空间的视角维度。

混凝土的楼梯有突兀的感觉，却因为吊顶而得到了化解。室内设计中固有的景观、街景的缺失也因为吊顶得到了弥补。空间变得柔化。

参数化的设计，根本在于一种创造，寻求一种外形，进而寻求视角上的实现。直线性的构件的密度、横截面、可控曲面的数量、半径及其旋转因为参数化的数码物件得到了变化。同时，自由流体的表面在美学的基础上实现了形态的变化

The owners of the small pension succeeded in purchasing the neighbouring building of an old family house of the countryside type. The commission involved designing and realization of the cafe with the bar "in the modern spirit" over an extremely short time and without the possibility of radical construction adaptations to the already reconstructed building.

The bar counter is situated at the front of the space providing the finest view of the area. The steel structure of the bar counter is clad with chipboard panels sealed and treated with paint. Some of the appliances are inserted under the bar and some of them are integrated into the side wall. The niches of the front wall serve for displaying both the drinks and glasses. The niches are illuminated with yellow neon lights, which carve up the black part of the interior like lasers. The high reflections make the space seem larger.

The primary function of the drop ceiling is to visually suppress the oddly rising piece of concrete staircase, which sticks out into the space, and to cover up the air-conditioning technology. At the same time the elaborate ceiling is meant to compensate for the missing view of the landscape or street with a view of "something interesting" in the interior, and last but not

least also to dampen the acoustics of the space.

The basic consideration behind the parametric design was to create an instrument which would not only serve to seek a shape and its testing in visualizations, but which would serve for the actual realization as well. This instrument is a parametric digital object which enables changes of his several parameters like: density and cross-section of linear components, number of control curves, radius of control curves, their rotation, etc. Parametric instrument enables morphology changes of the free-flowing surface based on the aesthetic demands.

B 流光溢彩 灯光设计
THE COLORS LIGHTING DESIGN

搜狐玻璃办公室
GLASS OFFICE

客户：搜狐网站
设计公司：AIM 建筑
设计师：温迪等
摄影师：杰里

Client: SOHO China, www.sohochina.com.
Design Company: AIM Architecture
Designer: Wendy Saunders, Vincent de Graaf, German Roig, Carter Chen and Jiao Yan
Photographer: Jerry Jin, SOHO China.

在上海，搜狐网站有一间新办公室。空间的基础设施以玻璃、镜材作为包裹。由于镜面的使用，建筑固有的高度与结构任由观赏。静坐于室内，抬头张望，便是上海的都市天际线。

膜状的天花吸引着对结构模块化设计的关注。光华漫射之间，各表面也在空间得到了反射。地板的铺设有的是岛上的石头，有的是地毯。由此产生的凝固般的感觉抵消着由于反射而导致的动感。

因为是租赁的空间，没有经过任何装饰。设计后的空间，目的在于强化客人的实际感受，同时给人一种富有灵感的奢华。

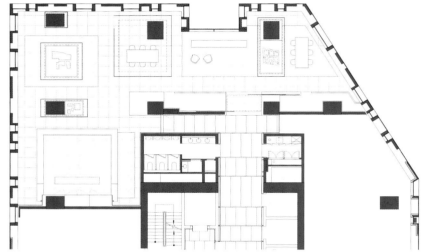

An all glass and mirror inner cladding exposes the infrastructure of SOHO's new office building in Shanghai. The glass creates manifold reflections of the sales models and meeting rooms, while leaving the original height and structure in view. This creates a "double reality" that merges with the stunning views of downtown Shanghai.

Membrane ceilings create extra attention for the models. Light and surfaces reflect throughout the space, even further diffused by half see through mirrors. Some of the floors are islands of stone or carpet, to create static moments to offset this sea of reflectivity.

As SOHO rents out the offices in this building in bare shell state, the main design idea is to show the customers what they are actually getting, and at the same time add a layer of inspiring luxury to it.

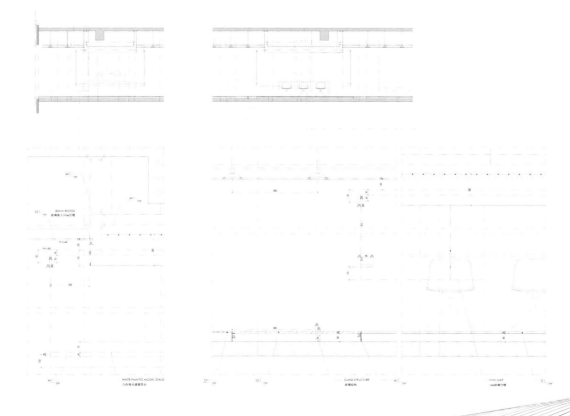

300 流光溢彩 灯光设计
THE COLORS LIGHTING DESIGN

室内灯光　其他
INTERIOR LIGHTING　OTHERS　301

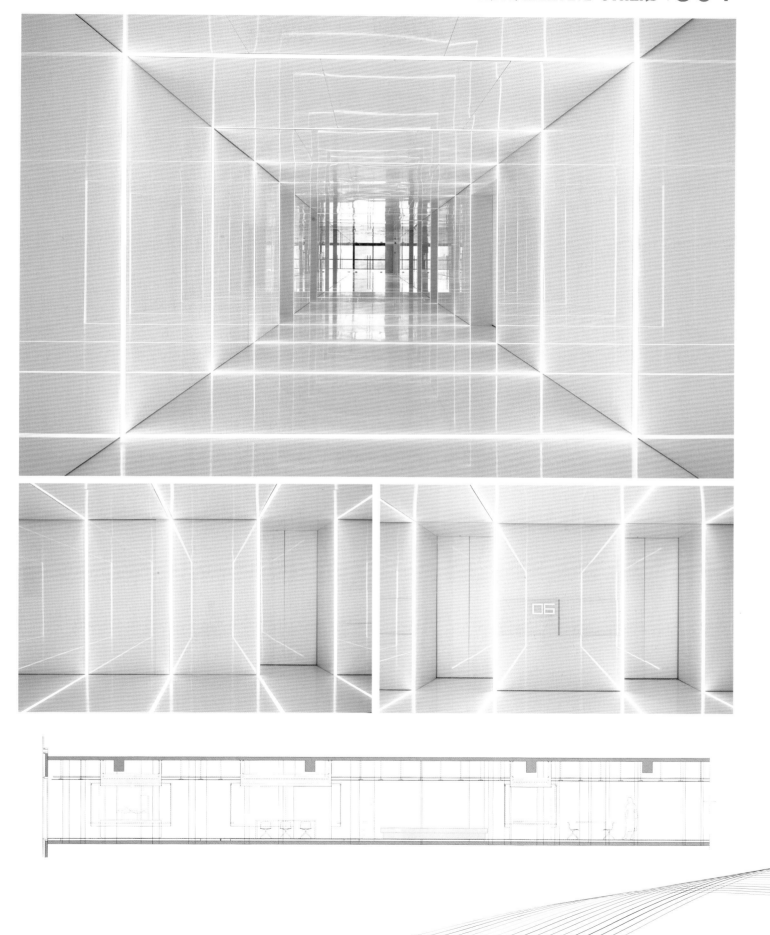

B 流光溢彩 灯光设计
THE COLORS LIGHTING DESIGN

梨泰院华丽丽会所
GLAM LOUNGE

设计公司：齐合作设计
设计师：申升勇、朴昌珉、朴赞未
主要材料：瓷砖、木地板、壁纸、面皮、彩钢板、铸板、油漆
面积：462.8 m²

Design Company: Chiho & Partners
Designer: Shin Seung-yong, Park Chang-min, Park Chan-un
Main Materials: Tile, Wood Flooring, Wallpaper, Color Steel Plate, Veneer, Backpainted Glass, Casting Panel, Paint
Area: 462.8 m²

梨泰院区华丽丽会所，分布于1、2两个楼层。爱尔兰风格的"健康酒吧"，传统欧式酒吧"华丽丽"，令人回想起美国20世纪二三十年代的"缪斯"会所。

"华丽丽"位于二楼，经典风格。接待室内设DJ操作台，如浴春风般的欢迎气质，吸引着客人入内一感身受。18个拱形窗，安于其中，创造着一种对称的内里。黑、黄铜色的古董家具，不同形状与质感，给人一种前卫的未来主义的感觉。

主酒吧天花的金属切割，三面弯曲、重叠，最大化地发挥着灯效与空间的视觉感。天花因此有了一种更加高耸的感觉，呼应着周围客人动感旋律的同时，升华着空间那种宏伟、戏剧般的张力。酒吧另一端，是一个独立的区域，布置有吧椅、桌台。

设计时，两层空间的差异得以充分运用。内里的环境便于客人之间相互交往。另一端的VIP区域是典型的韩国风格。东方的图案，洒下一地的神秘与梦幻。

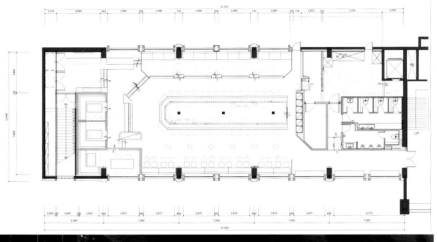

抬高的空间，俯瞰着置身于聚光灯下的人们。连接环形动线之间的通道是开放性的设计；玄关位于前面、侧面，化解了动线的承受力，便于客人往来出入。"华丽丽"酒吧对空间做到了深度利用，真真正正地体现了设计师的哲学：空间的能量之所以迸发，并非由本身引爆，多由出入往来者引发。

The project occupying the first and second floor includes Irish pub "Prost", traditional European lounge "Glam", and club "Mute" evoking of the 20s and 30s of the United States of America.

On the second floor is a classical lounge, Glam. The reception area combined with a DJ booth is inviting and immediately assimilates the visitors to the atmosphere. The designer installed eighteen arch windows to create a symmetrical interior, and the black, bronze gold, and antique furniture of diverse shapes and textures add an avant-garde and futuristic feel.

At the main bar, the metal upon the ceiling was cut, bended and overlapped on three sides to maximize the lighting and spatiality. Such an approach makes the ceiling seem higher as well as increasing the magnificent, dramatic feel in the space by reflecting the dynamic movement of the customers sitting around the bar. On either side of the bar are independent areas furnished with lounge chairs and standing tables.

The difference of floor levels creates an environment that encourages customers to socialize. The VIP zone on a side, on the other hand, is decorated with Korean characters and, more broadly, with Eastern patterns, emitting a mystic and dreamy ambience.

Its elevated floor overlooking the hall reflects those who wants to be under the spotlight. The functional efficiency was achieved by opening up the passages between circulation patterns, and locating entrances in the front and the rear side in order to disperse the overload of circulation and help the users to enter and move about the space more freely. The Glam lounge is characterized by the captivating space with depth under the designer's philosophy: A space should be set off by the energy who visits, not by the space itself.

B 流光溢彩 灯光设计
THE COLORS LIGHTING DESIGN

黑盒子
BLACK BOX-DOWLING BILLIARDS DARTS CLUB

设计公司：帕拉萨伊特工作室	Design Company: Parasite Studio
电暖设计：电子眼睛	Electric and Hvac Systems: Electric Eye
家具制作：GBT	Furniture Manufacturer: Galignum Bt
开发商：SRL会所	Developer: Club SRL
摄影师：马里斯 & 帕拉萨伊特工作室	Photographer: Dumitrascu Marius & Parasite Studio
面积：2 000 m²	Area: 2,000 m²

本案目的是在建筑最具地位的空间内，打造一个游戏场。一旦入内，人似乎进入到一种完全自我的天地，任由外界变化，沉溺于其中，却不自知。

设计的理念在于创造一个盒体式的空间，无论于其形状，于其理念，都可以实现与外界的切断，从而保持记忆永远在"里面"。于是，曾经的窗户，如今变成了一种陈列，类似于可以传递信息的显示屏，充当了向外界展示，与外界交流的工具。

2 000平方米的开阔空间，如若致人一种"凝聚"的感觉实是难事。更何况其原有功能布局混乱无章，大量的钢结构纵横交错。几经纠结，最终设计从"掩盖"到"漠视"，再到"利用"。通过整体设计，曾经的混乱借助于解构的形式，以投影的手法轻轻地飞舞在周围的墙体上。

围墙之内，是技术、功能分区与游戏空间的融合。或以动感、或以集中，或以各种配色在光照的辅助下，实现着不同空间的离散、整合。于是，有的空间用于游戏，有的用于酒吧，有的用于座位，有的用于公共区域。一切井然有序。

用于围护技术、辅助区域的墙体，因此也形成了此处空间的分割标志。有的专门用于投影，有的围护着抽烟区。有的围护着保龄球与游戏区。中心的立柱，却是唯一可以接触"世外"的媒介。顺着圆柱，任由光照倾泻。墙体上空的视频投影变化着其间的气氛。

家具铺陈，延续着空间设计的主题。各物件自有使命，自担角色。

自其始起，空间便尽显"流性"主题。图文并绘的地毯，无形中充当了空间的地图指示，方便客人在不同区域穿梭往来。黑色的背景下，自有强烈的色调，彰显着空间的标志。

整个内饰设计中，各元素如标志、家具等，无一不在强化、显示空间"黑盒子"的主题。

The brief of the project asked for the design of a games club located in one of the most significant spaces of the building. The main requirement was to obtain a space with a completely controlled atmosphere with no relation with the outside world in which you are not aware of the time passing.

The design proposal started from generating a completely black box, as well as a shape and as a concept, an abstract space that doesn't breathe towards the outside world but keeps locked within the memory of events. Therefore the existing windows

become showcases for exhibiting and communicating towards the outside, a sort of dynamic information screens.

The inside, although a vast in surface (2000 square meters) proved to be difficult to configure and especially to organize within a coherent unique concept, due to the existing random layout of technical areas and the crisscrossing massive steel structure. In the first design concepts we tried to cover and hide the existing structure, then tried to ignore it but came to realize that the best solution is to integrate the structure within the general interior design and to multiply it as a deformed projected image on the perimeter walls.

Over the actual structure we imposed two elements with the intention of unifying the interior—one is the continuous strip of the perimeter walls that wraps up the entire space, comprising gaming as well as technical areas, and the other one is the artificial lighting that through its dynamic, intensity and color scheme modifies the space and in the same time divides it discrete in different areas. The spaces that make up the interior are thus differentiated in dynamic areas for gaming, static areas for bar and seats, intermediary areas for communication, etc.

The technical areas and supportive functions for the club have been enclosed within the perimeter wall strip, forming punctual landmarks for the bar area, the circular projection wall, the smokers area, bowling or gaming area. The central cylinder is the only place where a filtered contact with the exterior is possible within the abstract interior—natural light enters at the base of the wall discretely indicating time's passing, while the video projections above the wall distort the interior atmosphere.

The entire furniture has been custom designed for this project, following the same conceptual design theme. Each piece of furniture serves its functional role and follows the concept.

The flux of movements in the interior space has been a major theme from the beginning on, and the carpet was personalized as an interior map that charts the different functional areas by the use of printed texts and paths. This adds to the fact that the entire space can be read as a communicating body that facilitates the movement within it and the reading of the separate areas. The strong colors chosen for highlighting the space set landmarks in the black surrounding.

For the whole interior design we strived to create a strong conceptual identity around the theme of the "black box", defining all the elements that make up the whole, the logo, the furniture ant the entire interior context.

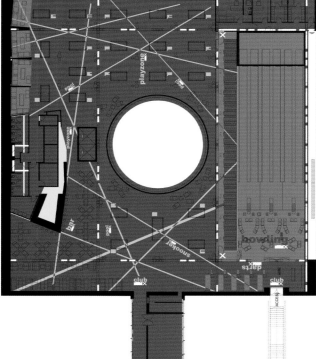

308 | 流光溢彩 灯光设计
THE COLORS LIGHTING DESIGN

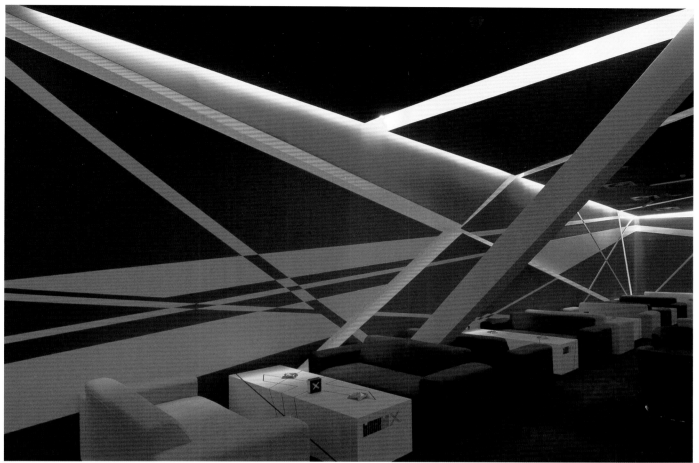
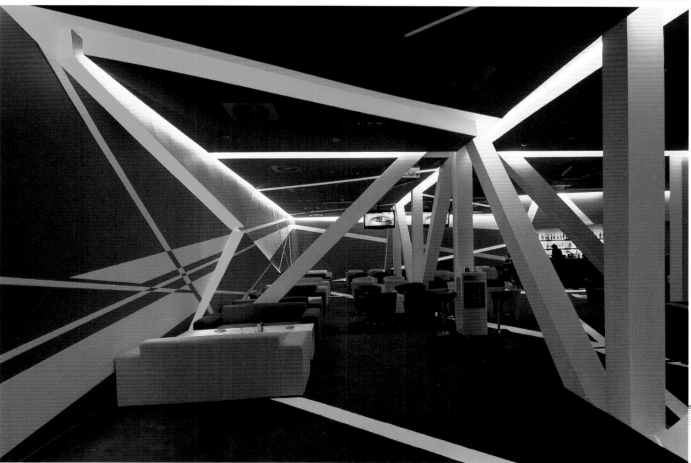

室内灯光　其他
INTERIOR LIGHTING　OTHERS　311

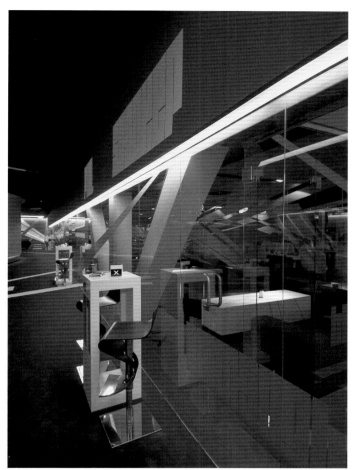

B 流光溢彩 灯光设计
THE COLORS LIGHTING DESIGN

克罗维克电影院
KRONVERK CINEMA

位置：波兰华沙
设计公司：罗伯特设计工作室
摄影师：安德烈
面积：1 045 m²

Location: Warszawa, Poland
Design Company: Robert Majkut Design Studio
Photographer: Andrey Cordelianu
Area: 1,045 m²

"克罗维克电影院"是俄罗斯电影院领军连锁影院。波兰华沙的这家分店室内设计由"罗伯特"工作室负责，旨在通过把影院"多元化"推广到一个新的高度。通过"连贯、持续"的方案，实现空间的组织及视觉上的身份识别。

本案空间的所有几何造型，如若寻踪溯源，都可以看到一定"直线"性的身影。这种理念实际来源于该电影连锁的商标，象征着与众不同的商标。线条经过繁衍、幻化，直线或成倍叠加，或交叉成形，组成不同模块。墙体、天花成普通矩阵，甚至LOGO都可以看作类同的表达。

内里不同空间持续运用着色度、几何形状，但每个区域各有区别，尽显编码式的美学特征。现代、几何形状的玄关，引领大堂。大堂具有典型的隧道特征，并以席座划分为不同的区域。大堂天花以灯光打造着辉煌的华彩，但却隐约表达着细致的图案。酒吧虽受限于规模，但却表达着一种俨俨的气氛。黄色的VIP酒吧中，单纯的色调中丰富的图案，表达着一种别致，一种能量。相反，VIP休息室，以黑色的平静、紫色的富贵、舒缓的基调，打造着一种亲密。其中，高档的用材，更是彰显着此处的别致。白色的廊道与黑色的放映大堂形成鲜明的对比。各空间至此，井井有条、界定明晰，无论功能、还是美学都融入了LOGO幻化而成的统一模式。

现代的模式，参照着俄罗斯的装饰。其中应用的模式、图块，是其他时代或传统装饰图案的现代诠释。设计定位完全符合客户的期望，并因其功能、现代化的解决方案、审美的一致性和连贯性以及"克罗维克电影院"品牌的认同，而受到客户的大加赞赏。

Kronverk Cinema is a leading cinema network on the Russian market. The concept developed by Robert Majkut Design to elevate the culture of multiplex cinemas to a very high level. It allows to be repeated both in new and existing cinemas as a coherent and consistent solution of the spatial organization and visual identification.

All spatial geometric forms present in this project are the consequence of a certain order of lines, which comes from the trademark—most of all from the distinctive symbol of crown. Derivation of these lines, their multiplication and crossing has created a flexible array of forms, which can be modified into various modules. In this way for each wall or ceiling there is a common matrix, one can find their counterpart in the elements of the logo.

Despite the consistent use of color and geometry in the interiors of Kronverk Cinema, each zone is characterized with a distinct and clearly codified aesthetics. The entrance to the cinema, modern and geometric, leads to the lobby area—a very characteristic tunnel, divided

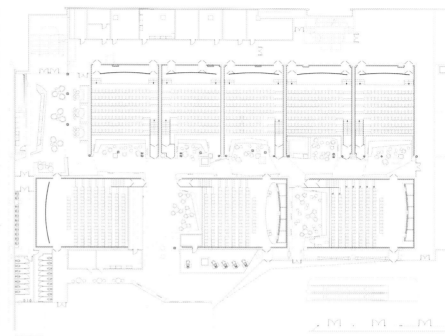

into spatial zones with seat modules. The ceiling of the lobby is a set for the magnificent play of light formed in a subtle pattern. Alcohol Bar, because of its scale, is to give the impression of saturated, patterned spaciousness. Completely monochromatic, yellow VIP Bar provokes a feeling of immersion into a unique and energizing space, decorated with figurative motifs. In opposition, the VIP Lounge, with its soothing tones of black and purple, creates a climate of intimacy, the feeling of being incognito in an elegant room decorated with precious materials. An unusual solution are the white corridors, strongly contrasting with the dark cinema halls. Thus, each room in this facility is clearly defined, both functionally and aesthetically—at the same time being part of a common pattern derived from the logo.

This design in a modern way refers to the Russian decorativeness. The patterns and forms applied in the project are contemporary interpretation of motifs known from other eras or traditional ornaments. The project fully met the client's expectations and was appreciated for its functionality and modern solutions, as well as for aesthetic consistency and coherence with the Kronverk brand's identity.

图书在版编目（CIP）数据

流光溢彩 灯光设计 / 黄滢 , 马勇 主编 . – 武汉 : 华中科技大学出版社 , 2015.4
ISBN 978-7-5680-0789-4

Ⅰ . ①流… Ⅱ . ①黄… ②马… Ⅲ . ①建筑照明 – 照明设计 Ⅳ . ① TU113.6

中国版本图书馆 CIP 数据核字（2015）第 073697 号

流光溢彩 灯光设计

黄滢 马勇 主编

出版发行：华中科技大学出版社（中国·武汉）
地　　址：武汉市武昌珞喻路 1037 号（邮编：430074）
出 版 人：阮海洪

责任编辑：岑千秀　　　　　　　　　　　　　　　　责任监印：张贵君
责任校对：熊纯　　　　　　　　　　　　　　　　　装帧设计：筑美空间

印　　刷：利丰雅高印刷（深圳）有限公司
开　　本：942 mm × 1264 mm　1/16
印　　张：20
字　　数：160 千字
版　　次：2015 年 6 月第 1 版 第 1 次印刷
定　　价：338.00 元（USD 67.99）

投稿热线：（020）36218949　　duanyy@hustp.com
本书若有印装质量问题，请向出版社营销中心调换
全国免费服务热线：400-6679-118 竭诚为您服务
版权所有　侵权必究